Praise for *Slick*

'The fossil fuel industry's campaign of disinformation and denial about global warming is perhaps the most consequential lie in human history. This comprehensive and carefully reported book shows how that lie was told in an Aussie accent, and why it mattered – and matters – so much.' — Bill McKibben, author of *The End of Nature*

'A diligent, urgent, trenchant study of political influence at its stealthiest by a journalist who always does the work.' — Gideon Haigh

'If you've ever wondered how the oil and gas industry was able to capture our politics and wreck the climate, this insider's guide is what you've been waiting for. *Slick* goes back to the beginning to show how these masters of deception have enmeshed Australia in an industry that has knowingly put all of us in harm's way. The fires and floods, the heatwaves and political shockwaves: they knew what they were doing, and they did it anyway. *Slick* reads like a crime scene investigation: how the oil and gas industry knowingly stole the future out from under us.' — Scott Ludlam

'*Slick* is an entertaining and often disturbing story of the scoundrels' gallery of oil industry lobbyists, shills and execs, together with their lap dogs in politics, who will not stop until the last drop of oil is extracted and burnt, and to hell with the consequences for the planet.' — Clive Hamilton

'Rich in science and history, Kurmelovs paints a vivid picture of how the ambitious "oilmen" of the '60s became the slick oil and gas lobby we know today, co-opting politics, media and science, all while knowing the risks to the climate. A must-read for anyone who wants to know how power works in Australia.' — Rachel Withers

Royce Kurmelovs is a journalist and writer whose work has been published by *Rolling Stone Australia, The Guardian, The Saturday Paper, VICE* and other publications. Royce's bestselling first book, *The Death of Holden*, was published in 2016, and his critically acclaimed fourth book, *Just Money*, was released in 2020.

SLICK

AUSTRALIA'S TOXIC RELATIONSHIP WITH BIG OIL

ROYCE KURMELOVS

First published 2024 by University of Queensland Press
PO Box 6042, St Lucia, Queensland 4067 Australia

University of Queensland Press (UQP) acknowledges the Traditional Owners
and their custodianship of the lands on which UQP operates. We pay our respects
to their Ancestors and their descendants, who continue cultural and spiritual
connections to Country. We recognise their valuable contributions to Australian
and global society.

uqp.com.au
reception@uqp.com.au

Cover design by Josh Durham (Design By Committee)
Typeset in 11.5/15 pt Bembo) by Post Pre-press Group, Brisbane
Printed in Australia by McPherson's Printing Group

  University of Queensland Press is assisted by
the Australian Government through Creative
Australia, its principal arts investment and
advisory body.

A catalogue record for this book is available from the National Library of Australia.

ISBN 978 0 7022 6860 1 (pbk)
ISBN 978 0 7022 6985 1 (epdf)
ISBN 978 0 7022 6986 8 (epub)

University of Queensland Press uses papers that are natural, renewable and
recyclable products made from wood grown in well-managed forests and other
controlled sources. The logging and manufacturing processes conform to the
environmental regulations of the country of origin.

'Australia's oil and gas industry is committed to and central to achieving net zero across the economy by 2050: supporting the transition away from coal, backing up renewables in electricity and enabling net zero fuels and technologies such as low-carbon hydrogen and carbon capture, utilisation and storage.'

— Samantha McCulloch, Chief Executive, Australian Energy Producers (formerly APPEA), full statement in response to multiple questions raised during research for this book, 22 September 2023

Contents

Introduction

Positive Energy for a Changing World

An eerie blue light washed the Brisbane Convention Centre ballroom: blue for petroleum, blue for burning gas, blue for blue hydrogen.

It was a Wednesday evening in mid-May 2022, day three of the oil and gas industry's big four-day national conference – not counting the golf tournament scheduled for that Friday.

The presence of the industry's executives, here on the banks of the Brisbane River, was one of life's great ironies. Two months ago, as the river rose under torrential rain, convention centre staff had worked through the night sandbagging the venue to keep the water at bay. The floodwaters had surged right across South East Queensland and down over the border into northern New South Wales, past Lismore and the Northern Rivers, into areas where the Black Summer bushfires had burned the year before. The water had yet to properly drain from Brisbane, but now venue staff were playing host to a crop of petroleum execs whose core business was driving a change in the chemical composition of the atmosphere that had made those floods worse.

The absolute cream of the industry had begun to gather about an hour before the doors opened. Those present represented some of the biggest oil and gas companies in the world – the ExxonMobils, the Chevrons, the Shells – as well as smaller players, public relations professionals, departmental bureaucrats and elected officials. They formed an impenetrable wall of grey suits in the foyer as they milled around, knocked back drinks and networked. Then, as the annual gala dinner was about to begin, the doors had opened and we were let loose inside.

My plane had touched down into the humid Brisbane autumn a few days before. This was postcard Australia, home to the Great Barrier Reef and the late, great Steve Irwin. A friend had picked me up from the airport and, on the way back to their place, given me a tour of where the floodwaters had risen. I hadn't been sure what to expect when I set out to cover the conference. The overall slogan they had chosen was 'positive energy for a changing world' – an attempt by the oil and gas industry to spin the pressure they were under as demand for action on climate change mounted. Over the next few days, I'd browsed the stalls showing off the industry's best efforts to suck more fossil fuel from the earth's crust, enjoyed the free lunches in the exhibition hall and been shamed by protesters alongside the other attendees when I turned up each morning.

It was sometime during the second day that the Australian Petroleum Production and Exploration Association's (APPEA) press guy asked if I wanted to come to dinner. We were in the media room – sponsored by Santos, one of Australia's biggest domestic oil and gas companies – when he caught me walking past. The media room was an air-conditioned refuge from Brisbane's humidity, well-stocked with fresh pastries, hot coffee and Santos-branded merch – all part of an evergreen strategy by governments and large corporates to curry favour with journalists. Time had even been carved out in the conference program for after-work cocktails and a meet-and-greet so the oil company executives could mingle with the nation's financial press.

The previous day, an email had gone out with an invitation to dinner, and the press guy asked whether I had seen it. I had – I just hadn't responded to it. Freebies are supposed to be frowned upon.

You could tell who the reporters with integrity were, because they left the Santos-branded drink bottles where they stood. And they were nice bottles.

As the press guy looked at me, I made a snap decision.

'Sure,' I said. 'I'll come to your party.'

His eyes widened; he seemed genuinely happy. Perhaps he was surprised. In the lead-up to the conference, I'd been cc'd into an email to another journalist who regularly covered climate change. It suggested to me that APPEA had run my background and lumped me in with the industry's 'critics'.

Ordinarily, tickets to the industry's annual dinner ran a punter $195. This came on top of the $1845 that APPEA members paid for standard general-entry tickets to the four-day conference, or the $2295 for everybody else.

The press guy handed me my free pass, saying it was supposed to be a hell of a show. It was a casual bit of sleaze, another well-worn strategy the organisation had developed back in the 1960s. Those who could afford the golden ticket bought themselves unfettered access to the nexus of Australian wealth and power, the part of the Venn diagram where government and industry overlap.

The room looked like a scene from *Blade Runner*. Once we were seated, wine flowed and filet mignon hit the tables as Senex CEO Ian Davies, APPEA chair, began his speech, telling his audience how the world looked 'a bit different' to when he'd last addressed the gala dinner crowd in 2019. Davies was referring to the upheaval wrought by the pandemic and the Russian invasion of Ukraine, but there were perhaps even bigger changes to come. We were just days out from the federal election in which Prime Minister Scott Morrison, who uncritically supported the gas industry, would lose handily.

'Through thick and thin, Queensland's oil and gas industry has persevered,' Davies said. 'In our industry, as in life, the *how* you do something is just as important as what you achieve.'

After platitudes came the awards. To help hand out trophies for 'environmental excellence' – trophies that the industry was awarding itself – Stuart Smith, the outgoing head of the National Offshore Petroleum Safety and Environmental Management Authority (NOPSEMA), was called to the stage.

NOPSEMA is the industry regulator; it's meant to be the cop on the beat, ensuring that oil and gas companies play by the rules. I gazed in amazement from a table at the back of the room, surrounded by APPEA's marketing team, as Smith – a tall, bland, bespectacled man – stood politely off to one side. As each name was called, he stepped forward for a photo, hands respectfully crossed in front of him. When the ceremony was over, Davies heaped effusive praise on Smith and encouraged the audience to applaud him for his eight years of service.

Later, when reporting out what happened at the gala, I asked NOPSEMA about this spectacle. In an email, a representative of the regulator said it had no issue with Smith appearing on stage during an industry dinner, as his contract was up in September. In NOPSEMA's view, the event offered an opportunity for his work in the role to be 'acknowledged' as he was winding up his tenure.

Eventually the awards came to an end, but there would be one more surprise. Through a blue haze on the main stage, the words 'Together We Shine' – the theme of the sold-out, 85-table dinner – were beamed onto two large screens.

This was a reference to the song 'Shine' by Vanessa Amorosi. It was an unfortunate choice. Whoever was responsible appeared to have latched onto Amorosi's hit because of its title without doing their due diligence. If they had, they would have found it contained lines like: 'You can give your life/you can lose your soul/you can bang your head/or you can drown in a hole/nothing lasts forever/but you can try.'

As it turned out, Amorosi had written the song as a teenager. She wrote the original version for a friend from high school who had taken their own life. She initially titled it 'Die', but her production team thought it too dark and it was later changed to 'Shine'. The original lyrics were: 'Nothing lasts forever/but you can try/look around you/everyone you see/everyone you know/is gonna die.'

Once desert hit the tables and the crowd was nicely toasted, the entertainment for the evening was announced: a private show by Vanessa Amorosi.

At this point in human history, it was hard not to conclude that the oil and gas industry was taking the piss.

Like millions of other Australians, I watched the east coast burn on the nightly news during the Black Summer bushfires and was stunned. The fire season began early in 2019, with the first blazes starting up around the Queensland border in June and then raging throughout southeast Australia for nearly a year. As the fires reached their peak in January 2020, the nation was transfixed by the images of a red sky over the small Victorian coastal town of Mallacoota as residents were evacuated from its beaches.

In response, the federal government was mute. Though it acknowledged the fires as a national tragedy, the prime minister and other leaders repeatedly spoke in their public statements about how Australia was a harsh land of fire and flood. The fires shouldn't be 'politicised' by talking about climate change, they said, especially while they were still burning. Their preference was to have that conversation later. Events showed it would barely be had at all.

As a reporter, I saw the devastation firsthand. I travelled to New South Wales while the fires were still smouldering and down to a scorched Kangaroo Island in their aftermath. I spoke to farmers broken by their horror at having to shoot their dying animals. Volunteer firefighters described their 'shell shock' after watching a pyrocumulus cloud build over a bluegum plantation until it collapsed like a bomb, showering fire over several square kilometres in a matter of minutes. At least 7.5 million hectares burned along the east coast during that fire season. Thirty-three people were killed, around 445 died indirectly from smoke inhalation and over a billion animals were thought to have perished. The embers were still glowing in some parts of the country as the pandemic erupted and the nationwide spread of Covid-19 forced cities to shut down in March 2020; in others, burned-out communities were just starting to assess the damage. Disasters aren't just more severe than they were in the past; they are more frequent, too – so perhaps we should have known there was more to come.

There followed a whole range of other disasters. In April 2021, Cyclone Seroja slammed into Kalbarri, Western Australia, and flattened the place. Six months later, giant hail shot holes through roofs, car windscreens and solar panels in Queensland's Mackay region. Then, early in 2022, the areas that had burned during the Black Summer bushfires drowned under torrential rain, when the Northern Rivers floods temporarily wiped towns like Lismore from the map. As the year rolled on, Victoria's Maribyrnong River flooded, and a wall built to protect the Flemington Racecourse in suburban Melbourne sent muddied water washing through homes on the opposite bank. In January 2023, the Fitzroy River in the Kimberley flooded, cutting off several remote towns for at least a week. Some described the flood – the worst in Western Australian history – as an 'inland sea'.

Over two years following the Black Summer bushfires, there were 11 declared insurance catastrophes in Australia, resulting in $13 billion in claims paid out. By 2050, the Insurance Council of Australia estimates that the cost of climate-related extreme weather events will hit $39.3 billion a year. This, for what it's worth, is just the projected cost to the insurance industry. It does not include the cost in human lives, the risk of ecosystem collapse or what happens when entire communities are displaced by a rising tide. According to the Internal Displacement Monitoring Centre, an international agency that tracks the number of people displaced by conflict, violence and disasters, an estimated 48,900 Australians were uprooted from their homes in 2021. Ordinarily, Australians only hear the phrase 'internally displaced persons' on evening news dispatches filed from countries in the developing world. The very notion that the country has its own population of internal refugees is inconceivable to the average person, let alone that these people were displaced by climate change.

It is easy to overlook these events as little more than isolated tragedies. Racked up alongside each other, though, they begin to show how a phenomenon too often spoken about in abstractions is punching through into the everyday.

Of course, the precise role climate change played in any one of these events is difficult to identify. What we do know is that climate change makes extreme weather events worse, and that the underlying cause driving the change is burning petroleum – or gas. Or coal. Whatever your fossil fuel of choice.

To appreciate the gravity of our situation requires a little context. It took 252 million years for the planet's oilfields to form. Across timescales impossible for the human imagination to grasp, algae and plankton fell to the bottom of warm, shallow oceans, where they became trapped, pressed and cooked by epochal geological processes. As they liquefied, they were transformed into raw crude, and the carbon they had consumed in life was buried, safely locked away for what should have been an eternity.

No-one is sure just how much oil humanity has consumed throughout two centuries of organised extraction. Burma became the first country to export oil in significant quantities when it shipped a supply to India in 1853; in the west, the first modern commercial well was drilled in Titusville, Pennsylvania, in 1859. Estimates of how much oil has been extracted since, vary wildly. They range from hundreds of billions of barrels to over a trillion. What matters is that burning fossil fuels today accounts for 86 per cent of the world's energy supply and around 75 per cent of current human-origin CO_2 emissions. In 200 years, less than a millionth of the time it took to create oil, humanity has busied itself releasing all that buried carbon back into the atmosphere – destabilising the climate in the process.

And we have long known about the problem. In 1856, American scientist Eunice Foote first described the power of carbon dioxide to trap heat and suggest this might alter the earth's climate. But, possibly because she was a woman, no-one took her seriously, and it was more than a century before she received credit for her work. Instead, the crown went to Irish physicist John Tyndall, when in 1861 he described the theoretical basis for the greenhouse effect and identified infrared light as the source of heating – going a step further than Foote. In 1896, Swedish chemist Svante Arrhenius published the earliest known paper predicting that burning coal would enhance the natural greenhouse effect, warming the atmosphere by a few degrees Celsius. This was a fringe idea at the time and would stay that way for another couple of decades. UK engineer Guy Callendar ran the first numbers confirming temperatures were rising in 1938 and presented them publicly the following year, but the idea would still take half a century to catch on.

If the general public were being kept in the dark and their political leaders weren't paying attention, the world's petroleum producers took it seriously enough. Running an oil company is not unlike gambling. Among the skills required by an oil or gas company exective, the most pivotal are risk management and long-range planning. As early investigations into city smog began in the US in the 1940s, fossil fuel executives followed the findings through to their logical conclusion and realised the implications. In response, they resolved to keep a watchful eye on the issue. The American Petroleum Institute, Exxon, Shell, BP, Total – all these organisations established some form of research program investigating the greenhouse effect, a few as early as the 1950s. But instead of alerting the public to the emerging problem they identified through their atmospheric research, they chose to conceal this information for decades – all the while claiming there was not enough proof to justify action.

One internal Exxon memo, obtained through a 2015 *Los Angeles Times* investigation, neatly summed up their strategy. Dated July 1988, it outlined the company's research on the greenhouse effect before concluding with the official Exxon position: 'Emphasize the uncertainty in scientific conclusions regarding the potential enhanced greenhouse effect. Urge a balanced scientific approach.'

And with that decision, an active disinformation campaign began.

There were other documents too – whole libraries of them – but it wasn't necessary to have done any homework to grasp what was happening at the Brisbane Convention Centre on the night of the gala. All that was needed was to pay attention to the names.

There was Coalition resources minister Keith Pitt, future Labor resources minister Madeleine King, former foreign ministers Julie Bishop and Stephen Smith, Queensland Liberal National Party MPs Pat Weir and Trevor Watts, and Labor MP James Madden – among others. Attempting to list all current and former political leaders at the dinner would have been an absurd and pointless task. What mattered was that they were here – a testament to the gravitational pull of oil and money. Even Greens leader Adam Bandt made an appearance at one point – outside the conference centre – to address those protesting the event.

Among the big shots in attendance were all the little people with no names: the teeth on the gears that kept the whole machine running, each responsible for a thousand individual decisions that added up to a greater whole. They were scientists, industry analysts, lawyers, engineers, consultants, staffers at regulatory authorities, service contractors, public relations specialists and marketing professionals. They were people with skin in the game, people whose income came from oil and gas in one way or another and whose nine-to-five enabled the continued extraction and combustion of fossil fuels.

The dinner's guest list suggested an uncomfortable degree of overlap between government and industry – a state of affairs some might call 'state capture'. Once a technical term used only by international relations theorists, the concept of state capture was popularised in Australia by former Greens senator Scott Ludlam in his 2021 book, *Full Circle*, alongside a campaign the Australian Democracy Network ran to challenge corporate influence on politics.

The World Bank defines state capture as 'the exercise of power by private actors – through control over resources, threat of violence, or other forms of influence – to shape policies or implementation in service of their narrow interest'. As an analogy, it is a helpful to think of government as a laptop; its institutions are the hardware and its policies the software. 'State capture' refers to a situation in which large corporations and business interests act as hackers, tinkering with institutional programming to generate their preferred outcome. Whether it's banana companies dictating government policy in

South America or preferential allocation of government contracts in the US, according to the World Bank the essence of state capture 'lies in a distinct network structure in which corrupt actors cluster around particular public organs and industries'.

My first attempt to write about state capture in Australia hadn't gone over so well. It was 2021, when, after the fires and the floods, there was a flurry of editorial interest in the concept. As people asked what caused the fires, they also began looking more closely at those who had been making decisions. Fingers were pointed at oil and gas producers, accusing them of lobbying against climate action – but always from a distance. No-one bothered to speak to them directly to ask what they thought about the accusation.

But I did – and I unknowingly did so just as the International Energy Agency (IEA) released a report saying that limiting global warming to 1.5° Celsius, as the world's nations had agreed at the UN Climate Change Conference in Paris in 2015, meant there could be no new oil, gas or coal investment beyond the end of that year. Looking for comment, I pulled an email address from an old APPEA press release and dashed off a series of questions without thinking too deeply about it. I was looking for their response to claims that their strategy of hiring former state and federal political advisers had created a 'revolving door' between industry and government and that this hiring strategy gave them undue influence over policy, particularly climate change policy. I also asked for a response to allegations that their relationship with key scientific institutions represented a conflict of interest.

To be fair, I was brusque. For me, it was just another job – I was obligated to give the industry a chance to respond to the allegations – and I had several other things to do that day. But for APPEA, it was personal. Within minutes of sending my request, I received a response:

Hi Royce

This reads like a complete stitch up and not balanced at all, so to get a response based in fact, can you please send me the following:

Who is making these claims and what exactly are they saying, what proof is offered to back up these stunning and defamatory allegations? e.g. I could say things about our opponents and their undue influence as well on politics and shareholders. What evidence do you have or are you just relying on misinformation from our opponents who won't even acknowledge the thousands of jobs we create, the heavy lifting we are doing in relation to climate change, the fact our industry was one of the first to commit to net zero emissions (so the govt may not have committed to it but we have, kinda ruins the angle of your story that we are stopping them doesn't it?) and the fact that without our industry all the things people hold dear such as computers, phones, clothes, etc., would not exist.

At least they said 'please'.

It was the first time I had encountered this kind of pushback. In my response, I explained that I could not ethically divulge my sources prior to publication. I pointed to material that had been published on state capture to show there was valid public interest in the subject. I also added that I was only trying to give APPEA an opportunity to respond.

The phone rang almost immediately after I sent my reply. The voice on the other end explained that hashing things out in a call would be 'easier' than 'generating a huge email chain' – which meant I wouldn't have a written record of our exchange beyond my notes. In short order, the caller heavily implied that APPEA was planning to sue various environmental and advocacy groups for producing a list of the former politicians, political advisers and government bureaucrats who now worked for the association, and that I'd be sued if I continued to write about the subject. They added that 'serious reporters' knew this was all nonsense anyway.

When I asked about the IEA report recommending no new oil, gas or coal investment, they shrugged it off.

'It's modelling,' they said. 'It's only one path to reach net zero.'

Rather than deterring me, this conversation only made me more curious about APPEA, its membership and its activities. The story ran.

A year later, sitting in the middle of APPEA's annual party, I looked around and thought about that interaction. At any other industry conference, this sort of wheeling and dealing between government and industry figures would have been unremarkable – even expected. But the people in this room were committed to the ongoing production and consumption of a substance driving climate change, which threatens the collapse of environmental systems on which humans depend on for our survival. Moreover, they had no intention of stopping. They wanted to grow.

It is not hyperbole to frame the risk in existential terms. In March 2023, the Intergovernmental Panel on Climate Change (IPCC) released one of the most significant publications humanity has ever produced. The Sixth Assessment Report summarised current scientific knowledge about climate change in one document, concluding that the effect of human activity on the atmosphere had grown more pronounced over the past ten years:

> Human-caused climate change is already affecting many weather and climate extremes in every region across the globe. Evidence of observed changes in extremes such as heatwaves, heavy precipitation, droughts, and tropical cyclones, and, in particular, their attribution to human influence, has strengthened ...

Among the crucial paragraphs contained in the report's summary for policymakers is a section outlining the reality of ongoing fossil fuel production:

> If the annual CO_2 emissions between 2020–2030 stayed, on average, at the same level as 2019, the resulting cumulative emissions would almost exhaust the remaining carbon budget for 1.5°C (50%), and deplete more than a third of the remaining carbon budget for 2°C (67%). Estimates of future CO_2 emissions from existing fossil fuel infrastructures without additional abatement already exceed the remaining carbon budget for limiting warming to 1.5°C (50%) (*high confidence*). Projected cumulative future CO_2 emissions over the lifetime of existing and planned fossil fuel infrastructure, if

historical operating patterns are maintained and without additional abatement, are approximately equal to the remaining carbon budget for limiting warming to 2°C with a likelihood of 83% (*high confidence*).

In plain English: the oil and gas industry must die for humanity to live.

This wasn't a political statement on the part of the IPCC. It was a statement of scientific fact. No tricky accounting method or corner-cutting could get around this physical reality. And yet fossil fuel producers seemed hell-bent on trying to negotiate with destiny. As an IPCC working group observed in 2022:

The interaction of politics, power and economics is central in explaining why countries with higher per-capita emissions, which logically have more opportunities to reduce emissions, in practice often take the opposite stance …

Reading that, I felt like they were describing Australia. On a per capita basis, Australia's emissions were nearly double those of China in 2022. Australia may be the twentieth-largest oil exporter, but until recently it vied with Qatar for status as the world's top exporter of liquified natural gas. As of December 2023, there were 30 major fossil fuel projects with a reasonable prospect of getting final approval to proceed – two whole years after the International Energy Agency warned there could be no more investment in coal, oil and gas if the world wanted to meet its climate goals According to a 2022 report by the US-based research organisation Global Energy Monitor, on paper Australia also had at least 600 kilometres of new pipelines under construction – and an oil pipeline was no short-run project. With an operational lifetime that spanned decades, once installed, the contracting parties were locked in.

Framed this way, Australia doesn't look like a country taking climate change seriously. Quite the opposite. It's almost as if we're doubling down – with the possibility of profound consequences.

Truth, I like to think, exists at one end of a continuum shaped like a horseshoe. At the other end are lies, and right down the bottom of the curve is bullshit. Where any given statement lands depends on its intent and deference to fact. Liars seek to conceal facts for fear of their implications, whereas bullshitters just don't care about them. Bullshitters are people who, in the words of American philosopher Harry G Frankfurt, author of the essay *On Bullshit*, do not reject the authority of the truth, but simply pay no attention to it at all.

'By virtue of this, bullshit is a greater enemy of the truth than lies are,' Frankfurt said.

I'm no expert, but it seems like you can place all forms of mistruth along this spectrum. For example, sharing an outdated news story without realising that new facts have come to light lacks intent. On the other hand, cherrypicking – the act of picking out certain details to support your point of view without regard for what the source material actually said – is a form of lie that falls higher up on the other side of the curve, as would paltering. 'Palter' is a long-forgotten term for lying by omission. For instance, if you walked into a second-hand car dealership and asked how the engine runs on a red 1997 Holden VT Commodore, the sales rep might tell you that he took that same car out on a two-day trip to the country just last weekend and had no trouble. Even if the car's exhaust blew a cloud of thick, black smoke the entire way there and back, if he made it home, he's not technically lying to you – just paltering with the truth.

What the conference had offered, over the past few days, was a tasting plate of different kinds of misinformation: a dollop of mistruth, a heaping of palter and a smattering of bullshit. Up on the main stage, the industry had sought to project a carefully curated image of itself as bold, certain and in control of its future – but looming over the proceedings was the work the IPCC was doing to galvanise action on climate change and the recent IEA report that shocked the industry.

APPEA chair Ian Davies had not acknowledged the IPCC's work in his opening address to the conference, but the IEA report had rated an honourable mention.

'Under the IEA's net zero scenario, the world would still be using oil and gas in 2050,' he said. 'So, if we're serious about decarbonisation, we have to focus on reducing emissions from their production and use. This puts our industry at the frontline of the decarbonisation challenge.'

What Davies said was technically correct but, divorced from context, made it seem like the IEA had endorsed a thriving gas industry out to 2050. Listening to him, you would never have known that the IEA had called for an end to all new oil, gas and coal investment. As a salesman, he was so smooth that, by the end, I almost felt an urge to rush out and buy shares in Exxon.

Almost.

Away from the glitz of the stage lights and the television cameras, the technical presentations had offered something closer to the truth about what the industry was really thinking. These presentations had been run four at a time, in separate meeting rooms, away from the main stage. Subjects included everything from highly technical discussions about the latest drilling techniques to speculation about the future of hydrogen.

One talk had focused on decommissioning old oil assets – a hot issue for an industry confronting the prospect of its own demise. Another offered suggestions about how companies might better attract millennials into their workforce, as they were increasingly finding no-one wanted to work for them. A third discussed the latest research on handling addiction and suicide among oil rig workers – a massive problem in the industry, and one of its 'dark secrets'.

Perhaps the only truly honest people at the conference were the industry's lawyers. There were two separate presentations covering the legal risks associated with climate change and offering free 'practical tips' for handling potential climate litigation. In one talk, these included pointers such as 'Get to know your opponent', to understand how the lawsuit was being funded, and 'Get to know your judge', to 'learn a bit about how they've previously dealt with class actions'. At the conclusion, the lawyer warned her listeners: 'We're going to see more of these actions.' Neither session, however, was well attended.

The spectacle of a senior partner at a law firm talking to a half-empty room of oil and gas people about the prospect of being sued for cooking the planet suggested something about the overall vibe.

It was a feeling replicated across the city at that time. Having picked me up from the airport, my friend stopped for petrol. As she pulled the car to a stop, the price at the pump was eye-level. Unleaded was $2.24 a litre that day and Brisbane had turned into a sauna with successive days of 100 per cent humidity. As the oil industry was getting richer off my friend, she explained how she and her partner had spent the other day scraping the mould off the walls of their laundry and carport. It was a futile effort. Every surface seemed to sweat, creating the perfect conditions for thick sheets of black mould to reappear overnight. Bleaching it off walls and surfaces did no good. The next morning it was back again. The World Health Organization, for what it's worth, says there is no safe level of exposure to black mould.

The Brisbane Convention Centre was a refuge from the sweltering heat outside, but try as they might not even the titans of the oil and gas industry could escape their environment. The Australian Oilfield Golf Tournament – set up in 1966 as a way to break the monotony of the event – had been scheduled for the Friday but would end up cancelled owing to the torrential rain.

The competition's chair, Kylie Carre, lead analyst with Shell subsidiary QGC, sent around a note explaining with 'great regret' that the event would not go ahead. The Brookwater Golf and Country Club had flooded with the rest of the city, forcing it 'closed for a week and still not drained'.

'As most are aware there has been a major weather event over the last week in the greater Brisbane area, significant amounts of rain,' Carre said. 'Brookwater have advised that they will no longer be able to accommodate the AOGT for 2022.'

I was collecting my bags from the coat-check after the final panel when I overheard them: two tall white men in blue suits talking about climate change. They were part of the corporate hierarchy of one company or another, and one of them was getting really worked up about the idea of achieving net zero by 2050.

'It's bullshit,' he said. 'It's all bullshit.'

Among the closing events had been a panel discussion which questioned the idea that decarbonisation was practical or even feasible. Governments couldn't have it both ways, the panellists said – a point which the man was in furious agreement with. The world was sending contradictory signals, he complained: demanding a fossil fuel phase-out at the same time that governments were telling oil and gas companies to produce more in response to a supply crunch following Russia's invasion of Ukraine.

I'd spent the last three days listening to a whole lot of guff involving some variation of the idea that industry 'understands 100 per cent our obligations to decarbonise'.

These men, when they were being honest, simply didn't believe it was necessary for them to do anything about climate change, so long as the world remained hooked on oil.

How could they not? We were in an industrial-grade echo chamber. They had devoted their lives to an industry that posed an existential threat to humanity and had to end, but their salaries depended on it continuing to exist.

It was a joke, and yet speaking to those in the environmental movement afterwards, I was struck by how they tended to talk about petroleum companies and their industry associations as if they were a monolith – a crop of slick, super-sophisticated lobbyists who were very good at being bad. Oil and gas executives spoke about the environmental movement in similar terms. Each considered the other a 12-foot-tall giant. Based on everything I'd seen, these caricatures didn't exactly check out.

This started me thinking more closely about state capture, influence and climate accountability – ideas which form the basis of this book. In my reporting, I had already investigated how oil and gas companies had worked their way into our schools, our universities, our sports clubs, the arts and the public service. Others had written about how they have shaped our government's response to climate change.

What I wanted to understand was how it all fit together. Specifically, I wanted to know what Australia's oil and gas industry knew about climate change, when they had first known it, what these companies and their industry associations had done with this knowledge, how they had coordinated with their international counterparts, and how they had managed to insert themselves into so many aspects of public life. In short, I wanted to understand how this influence actually worked.

The result is this book about Australia, where oil and gas companies have come to dominate everything.

EARLY PERIOD

EARLY PERIOD

1

You Always Have the Roof

It's been said that the lived experience of climate change is watching videos of terrible things happening to other people on your phone until it inevitably happens to you. This is true, but in between those big, dramatic moments is the day-to-day accumulation of subtle inconveniences that eventually add up to make a place unliveable.

The year was 2022, and it was Kate Stroud's second big flood in five years when 'the bad thing' happened to her. She first noticed the cow a little after dawn. Her partner saw it first – or rather, heard it: the sound of something caught in a tree on the front verge, thrashing about between the pounding rain and the gurgling rush of floodwater.

'Is something alive out there?' he asked.

'I hope not,' Kate said.

Looking out through her bedroom window at the park that had now become a lake, she scanned the body of water. When the sun set yesterday, the field had been green and lush. Then, at 3 am, a text message came saying the levee had broken. Two-and-a-half hours later, the milkshake-brown water bubbled up from beneath the floor and through her carpet. According to the council maps, a flood would have to reach 13.3 metres before it came through the floor of her raised Queenslander in this part of South Lismore.

By the time they noticed the thrashing figure outside, the water in their home was waist-deep and freezing cold. From the window, they watched the shape caught in the tree dislodge and be swept towards them before the current pinned it to the railing of their deck. Wading out to investigate, they saw the creature – a cow struggling to keep its head above water. The animal was four times Kate's size, and the first thing she saw was its eye, looking directly at her.

'It had this panic in its eye,' she says. 'It was saying, "Help me, help me. You need to help me."'

Kate froze. Everything about this moment was incredible. Irrational. Preposterous. Whatever word you wanted to use, it still wouldn't capture the feeling of standing in the middle of something so far outside normal that it upended your understanding of the world. It was Monday morning. She should have been going to work.

'Holy shit,' she said.

As she stared, her partner rushed by. He'd read the situation and wanted to help. If the 300-kilogram animal stayed like this, he reasoned, it might become stuck beneath the bullnose ceiling of their deck. Its chances of survival were already low, but it certainly wouldn't have any chance if it stayed where it was. He gently pushed the cow to free it, and they watched it wash helplessly down the street, carried off by the floodwater. Kate hoped it would be okay, even if something inside her didn't really believe that it would. She couldn't help think it a symbol of everything happening around them: the rain, the flood, all the people they knew right across Lismore waking up to catastrophe. She hoped they would be all right.

The year Kate was born, humanity pumped 20.61 billion tonnes of CO_2 into the planet's atmosphere. It was 1986, the year of the first big international meeting on climate change – the Villach conference. In those days, Kate's family lived in Bayswater, a suburb six kilometres northeast of Perth's CBD, in a house built in 1960 by her grandparents.

The land around these parts belonged to the Mooro people of the Whadjuk Noongar nation. In 1830, following the arrival of Europeans, the first attempt to colonise what is now known as Bayswater failed when flooding, unusual in the area, wiped out crops and homes, forcing any colonial who didn't die to leave. When Europeans returned in 1885 to try again, they built a shantytown of humpies – small, temporary shelters – on the banks of the Swan River. With time, a railway was built and development followed, but Bayswater never fully shook off its reputation as an isolated spot at Perth's city limits.

It may have been a sleepy outer suburb by the time Kate came into the world, but it was home. She spent the first 16 years of life in this world, and when she finished high school, earlier than most, she found herself caught in an awkward liminal phase. She didn't know what to do next, so she played for time, studying at TAFE and travelling.

It was around this time that Liberal prime minister John Howard announced that his government would not ratify the Kyoto Protocol, an international treaty that set national targets for the limitation of greenhouse gas emissions. He had waited for World Environment Day to make his intentions clear. Standing on the floor of parliament, he declared Australia would go it alone on climate change.

'What the Labor Party wants us to do is sign up to something that will place burdens on Australian industry, but not impose the same burdens, Mr Speaker, on industries of other countries which could well be our competitors,' he said.

'For us to sign the Kyoto Protocol in its current form would destroy jobs in many of the industrial areas in Australia – it would be bad news for the Hunter Valley region, it would be bad news for the electorate represented by the member for Hunter, it would be bad news for Australian exporters.

'Overall, I'm not going to be a party to something that destroys jobs and destroys the competitiveness of Australian industry.'

The year was 2002. Whether Kate was paying attention or not, the decisions taken at this time would profoundly influence her future and livelihood. Howard's words would define and entrench a political consensus, allowing those with the power to do something about climate change to wash their hands of the issue. Half a decade would go up in smoke before Canberra was dragged to action. In 2007, when Labor's Kevin Rudd won government, he ratified the Kyoto Protocol on his first day in office. Climate change, Rudd said, represented 'the great moral challenge of our time'. Over the course of his prime ministership, and that of Julian Gillard, Labor would go on to create the Climate Change Commission, the Climate Change Authority, the Clean Energy Finance Corporation, and set up an emissions trading scheme. Every single one of these modest reforms would be unpicked and unwound by their successors in the Coalition.

In the meantime, as all this was unfolding, Kate had applied to Southern Cross University in Lismore to study a degree in contemporary music. Acceptance meant moving, so when confirmation came, she began looking for a place to live.

Lismore seemed as good as any other place to be. It was a small rural town with not much going on, or so she thought. Kate reasoned she might actually get something done without the distractions of city life. She would eat those words upon discovering its live music scene.

On moving day, she packed her life into her RAV4, and her ex-boyfriend offered to come with her. The drive took a week, and they slept in swags under the stars. Driving into Lismore for the first time, the roads were peppered with potholes, and the cane fields opened up into flat, empty vistas along the flood plain. Kate was trading the dry Mediterranean climate of Perth for the subtropical climes of the Northern Rivers, where the air smelled deep, earthy and wet.

The year Kate turned 29 and finished her degree Tony Abbott was entering his second year in office after crushing Rudd at the 2013 federal election. During an appearance on ABC *Insiders* in September 2013, Abbott insisted he did not deny the science of climate change. 'I think that climate change is real [and] humanity makes a contribution. It's important to take strong and effective action against it,' he said.

But Abbott's actions spoke for him. During his first month as prime minister, he had gone on a tear. His government closed the Climate Commission, promised to kill the Climate Change Authority and began drafting legislation to repeal the Emissions Trading Scheme. Later, they ordered the Clean Energy Finance Corporation to stop making new investments.

To Kate, a world away in Lismore, these events were troubling, but remote. The implications didn't extend beyond the end of the nightly news. For someone just getting her life together, there was no way to appreciate how decisions being made somewhere else – the latest in a chain of thousands that had already been made before she was born – might yet become personal.

And life seemed so full of opportunity. She was working with local music festivals and had grown an open mic night from nothing. She started organising rooftop gigs above her apartment, too, attracting big-time artists, until the property owners found out and limited capacity to six at a time.

Kate's rooftop gigs were so successful that she and a few other people began talking about going legitimate and opening a commercial venue of their own. Before long, it was a reality: they found a shopfront on Woodlark Street, in the busiest part of town, with a three-storey warehouse out the back they named the 'Dusty Attic'. Over eight months, they collected beautiful, deeply coloured vintage velvet couches, bought sound equipment and tools. They did as much work on the fit-out themselves as possible, but they also took on debt to finance it. When the approval process stalled, they took on more debt to cover the rent. And then the 2017 flood happened.

What no-one on her crew realised was that the Dusty Attic lay at the lowest-lying point of the Lismore CBD. Kate was working at her second job, a sign-writing business she operated out of the venue, on the morning the water rose. It had been raining constantly for days when she first checked the Bureau of Meteorology radar, wondering when it might end. The ink-blue shapes settled over the map, but she didn't really understand the significance of what she was looking at. For Kate, the historic flood markers all over town were just part of the streetscape; an interesting historical artefact with no clear application to the present.

By about 3 pm, the water had crept up through the Browns Creek car park and pooled on the road outside. The 'oh shit' moment came two-and-a-half hours later as it licked the front step of the shop, threatening everything they had built. They worked frantically, sloshing through the rising tide, to remove what they could. Anything that couldn't be hauled away was raised up onto the second-storey mezzanine to keep it dry. It wouldn't be enough, though. By 6.30 pm, there was enough water to paddle a canoe through the shop. Come 9 pm, when she left the building, it was already flooded, the water chest-deep.

At its peak the next day, 31 March, the floodwater hit 11.59 metres. Kate took it all in from the window of her second-floor flat. Her housemate, a solicitor and Lismore local, was out of town in Byron for a training course when it happened. Kate messaged another friend who lived over in North Lismore about what she should be doing. This friend was another Lismore local who knew flood. She told Kate that she'd be fine. The flat had running water, and she had supplies.

'Worst-case scenario, you always have the roof,' her text message said.

Kate lived on the second floor, above a shop. The shop was underwater, but she was safe, so she waited out the flood in her apartment. It was a life-changing experience, she recalls. She opened the window and sat on the awning, gazing out at the lake that lay where a town once stood. The muddy water was visibly slicked with rainbows of oil – an indicator of a major health hazard. With water in streets, canals formed, lined by the upper storeys of storefronts, and Lismore looked like Venice, only eerie and still. With the exception of the occasional friend kayaking past to fetch supplies, humanity had retreated from this place.

'Time had stopped and warped, and all the to-do lists that were just so important two days ago had drifted down the river into oblivion,' Kate says. 'It was a real perspective changer.'

The day of the flood, her housemate arrived on a kayak with

supplies of fresh fruit, vegetables, cheese and wine. She moored at the awning, and they sat for a while before she informed Kate of her mission. There were important legal documents back at her office, she explained, and she had to rescue them. Kate offered to join her; in the rush to get home, she had left her laptop on the third floor at the Dusty Attic. It contained her life. Losing it meant losing everything. She had to rescue it.

As they paddled out, the city was still. When they arrived at her friend's office, the water had started to retreat, but the alarms were going off. 'Floods are a chorus of alarms,' Kate would later say. Emergency lighting lit up the hallways in cool white light. Moving through it felt like something out of a movie. 'Like the zombies were going to jump out at any moment,' Kate says.

Having successfully retrieved the documents, they paddled back around to the Dusty Attic. By then the water had dropped half a metre, so they kayaked in through the second-storey front window to the rear of the warehouse and then climbed out onto the mezzanine platform, where a ladder led to the office on the third floor. Looking around, Kate noticed the toppled sound gear and her vintage couches floating on the water's surface. All the furniture was covered in thick, wet mud. That was the moment she first understood the extent of the damage.

It was an experience that would nearly ruin her financially. As the Dusty Attic hadn't technically been trading at the time of the flood, the business didn't qualify for government support. The venue managed to open eventually, but the business never fully recovered. It struggled on for another couple of years, until the pandemic hit and it finally shut up for good. Kate would look back on the closure as a relief, in a way, and for a time she settled into a new routine, believing the worst period of her life had passed.

Nearly two hundred years ago, when Lismore was first established, the local Bundjalung people tried to warn the early colonists they were making a mistake. Annual rainfall in the area tips 1600 millimetres, three times the average across the rest of Australia, and the Bundjalung had inherited stories from their ancestors of vast floods turning the entire basin into an inland sea. During heavy rains, the water flows off the surrounding hills into creeks that feed the Wilsons River and Leycester Creek. Smack bang in the middle, right at the point where they meet, lies Lismore.

Nobody listened to the Bundjalung. The ancient combination of greed and racism meant the early colonists shrugged and said, 'She'll be right.' At that time, the whole area was covered by subtropical rainforest, known by the colonists as the 'Big Scrub'. Timber cutters moving upriver from Ballina on the coast in search of Australian red cedar – known as 'red gold' – realised they could use the area that became known as Lismore as a launching point for their hunt. To bring the logs back to town, they would float them down the river on barges. The desire for profit overrode common sense, and the early town grew up around an inland port.

Lismore has flooded 150 times since 1850. In that period, the nature of those floods has changed. Throughout a century and a half of colonial occupation, the town has sprawled out across the flood plain. Deforestation has changed the topography and the rivers themselves; when it floods now, the water moves faster, over open country, and hits with more force. Any debris swept up along the way moves with enough velocity to punch holes right through the walls of the district's old weatherboard homes.

The size of these floods is determined way out over the Pacific. When the equatorial trade winds grow stronger than usual, they push the warm water and warm, moist air at the surface of the ocean west, towards Asia and Australia. To replace that warm water, deep, cool water is churned up in the central and eastern Pacific, off the coast of the Americas. The result is La Niña, a climatic cycle that brings torrents of rain to much of northern and eastern Australia.

When those same winds grow weak, or even reverse, the warm water and warm, moist air are pushed back east, towards the Americas, and ocean temperatures in northern Australia cool, in an event known as El Niño. In an El Niño season, rains that would otherwise fall across northern and eastern Australia fall on the central and eastern Pacific instead. The lack of moisture over the east coast of Australia brings drought and baking heatwaves – a mix that creates conditions like those which led to the 2019/2020 Black Summer bushfires.

When a La Niña event coincides with a negative Indian Ocean Dipole – as occurred at the time Lismore flooded – these floods are even bigger. A negative Indian Ocean Dipole (IOD) is a similar event to La Niña, just on the other side of the country. A positive IOD means weaker winds, like El Niño, and a negative IOD means stronger winds, like La Niña. When a La Niña event coincides with a negative IOD – as occurred at the time Lismore flooded – the combination can bring huge amounts of rain to southern and eastern Australia, making floods even more dramatic.

Thrown into this mix is climate change. Since the 50-year period from 1850 to 1900, the earth has warmed by 1.09° Celsius, with most of that heat absorbed by the oceans. Warmer ocean temperatures increase the amount of evaporation, which warms the atmosphere as it is absorbed, causing it to hold more moisture. When the rain comes, it falls all at once, in devastating quantities.

Two days before the 2022 flood, Kate Stroud had a feeling something bad was coming. She watched the storm system settle over Lismore on the weather app, and this time she knew what it might mean. She had a 'visceral' feeling that things weren't quite right – that nature was angry. Ever since she had moved to the community, she had heard the old wisdom that flood was simply a fact of life in Lismore. But this seemed like something else. As the weather turned, it rained hard and then harder.

'It got really wet, really quick,' Kate says. 'We were watching this weather pattern just continue and stay and stay and stay.'

On Saturday night, she and her partner had gone to a theatre show – a political satire – in the centre of town. When they woke on Sunday morning, word had already gone around that things were 'feeling floody'. People were advised to prepare. They spent that day gathering supplies, moving their cars to higher ground and preparing the yard. Anything loose had to be tied down or removed onto their deck. Bins had to be secured; oils and other chemicals had to be properly stored. Anything valuable was moved to a friend's place with no history of flooding.

It wouldn't be enough – not that Kate knew it at the time. Privately, she wondered whether it was all even necessary. Their South Lismore home was supposed to be safe. She and her partner had specifically bought the raised Queenslander – a house built on stilts to allow floodwater to pass harmlessly beneath – after watching the 2017 flood swallow the Dusty Attic. The worst-case scenario – the 1974 flood, the worst then on record – had peaked in Lismore at 12.11 metres. Even if the water was that high, it should sweep right underneath the house with a metre to spare.

Knowing this, the couple focused all their effort outside their home. When they were done with their own yard, they offered to help others. There was a sense of solidarity in those hours. People rallied, offering reassurance and assistance. Neighbours helped neighbours as they conferred over the fence and traded free advice. Looking back, Kate says it was 'humanity stripped bare', a cohesion arising from a mutual instinct for survival.

As the sun fell over Lismore that evening, there was a sense that, come what may, they were prepared – that everything they needed to do had now been done and all that was left was to wait out the hard part. They would be wrong.

There would be no calm on that fateful Sunday night. Knowing something would happen was one thing; it was another not knowing how bad it would get. As the hours wore on and tension built, sleep was nearly impossible. Outside, the rain hit like a bomb, pounding away at their roof. A rainfall gauge at nearby Goolmangar Creek would record 531 millimetres of rain that night, just under half the average rainfall Sydney gets in a year. Text messages flew back and forth as people monitored what sources of information they could – the Bureau of Meteorology, the Lismore City Council, the State Emergency Service (SES), Facebook groups, and friends who had experienced floods or lived close to the river.

At around 11.30 pm, there was a knock on the door. A heavy-set, middle-aged man from the volunteer SES wanted to speak with

Kate and her partner. He wore a waterproof rain jacket with a hood. Behind him, the emergency lights spun on the SES truck, cutting through the rain. Kate thought he seemed calm as he explained he was just making sure they knew a flood was coming. There was definitely water already in their yard, he said, but it wasn't very high.

'The levee's going to top at 6 am, but you should expect water at your gate by about midday tomorrow,' he said.

The man asked if they had supplies. They did. He asked if they were going to stay. They were. He wished them luck, and then, as he left, he tied yellow SES tape to their front gate as a signal to others that there were people in the house. Kate closed the door, went to the bedroom and tried to get some rest. Just after midnight, an update came through: the timeline had been brought forward. The levee would overtop at 3 am. The predicted water level had jumped three metres in a matter of minutes – and that's when a new wave of text messages started to land.

'Are you going to evacuate?' they asked.

'You should get out of there,' people said.

But it was too late to evacuate. The couple answered the texts, saying they were fine, they were staying put. There was no point leaving when there were others who needed those beds in the evacuation centres more, anyway. Her partner tried to grab a few hours' sleep on the couch, and Kate lay down on the bed again.

At 3 am, another update came: the water had breached the levee. As she was reading it, Kate heard a crash downstairs as a cabinet in the garage toppled over. It was the first sign something was deeply wrong.

'That's when our day started,' Kate says.

She woke her partner and brought him up to date. For the next hour, they worked. They moved food, blankets, fresh water and power tools into their roof, then tried to get their possessions up off the floor, but no matter how high they moved them, the water would soon rise within reach. Outside, the morning sky was still an inky black, but from the back door they watched as the murky water swallowed their stairs. That's when Kate had a text message from 'a dear friend' – her guide from North Lismore during the last flood.

'I'm so sorry,' it read. 'The water is going to come into your house. You need to pack a backpack with water, food, warm clothes, charge your devices and if you need to, get into your roof. I'm sorry to sound scary. I love you.'

The cow washed away as swiftly as it had crashed through their morning, just after dawn. Standing waist-deep in the freezing water on the back deck, the couple waded back inside, looked around at their possessions and realised trying to save them was pointless. Together, they climbed up into the cavity of their roof, and Kate lay down to catch her breath. Their next-door neighbours were also in trouble. Normally, they could hear each other call out from their decks, but the rapping of the rain was so loud that they had to shout to each other over the phone. Their neighbours had two small children and briefly considered coming across to Kate's place so they could all shelter together. When they looked out at the water, a raging torrent was running between the two houses where the driveway once was, so they stayed put. Instead, they made a pact that if either family managed to get a rescue boat, they'd make sure the others were also rescued.

No help would come for the first six or so hours they remained in the roof. Their neighbours cut a hole in their roof and managed to flag down a passing rescue boat from the 'Tinnie Army' – a spontaneous volunteer navy that had begun staging rescues when the authorities were overwhelmed. Kate's neighbours were already on the boat, pulling away from the house, when they called to say there was no room. They promised to send the boat right back once they were safe. In the background, Kate could hear their three-year-old daughter yelling, hysterically, 'That's my house! That's my house!'

At that point, the couple decided it was time to move. Kate grabbed the backpack she had prepared and climbed back down into the water. It was now neck high. As she stood on the kitchen countertop, she watched her belongings float around on what looked like chocolate soup. As she looked for an exit, she peered out through the kitchen window and watched a pair of her shoes drift away. The flood had forced open the French doors in their bedroom and they were losing their things.

'We've sprung a leak,' she called out to her partner. 'My favourite shoes are floating off down Casino Street!'

It was then she noticed the jet ski moving between their house and their neighbours'. Atop was a burly man in a black wetsuit – 'He looked very GI Joe,' Kate says. Thinking fast, she banged on the glass to get his attention. Hearing the thuds amid the drumbeat of the rain, the man scanned his surrounds and saw Kate's face through the window. He dipped in towards her, and Kate and her partner took their chance to climb out.

As she left, Kate closed the window behind her. It was a meaningless gesture in the scheme of things, but she didn't want to lose anything else. Once on board the three-seater jet ski, they rode down Casino Street, up Union and then alongside the bridge – or where the bridge was before it had been submerged. Cutting across the raging flow of the river, Kate shut her eyes and held on as the man on the jet ski accelerated onto Woodlark Street. It was the same route Kate would normally take to get to work, only this time she had to duck live powerlines. As they pulled into town, they had a chance to witness the devastation. Kate pointed to an awning sign she had recently painted, now underwater.

'I had to get a lifter to paint that,' she said.

On 28 February 2022, I had just started my Monday-morning shift when my editor assigned me a story about Lismore. Apparently, there had been a flood.

The first person I called never picked up. I learned why when the second call I made was to Sue Higginson, a future New South Wales Greens MLC. Higginson had evacuated the night before because she'd had a bad feeling. When I reached her, she almost yelled down the phone about the unfolding catastrophe.

'There's screaming, traumatised people on rooftops,' she said. 'My friends are out there right now, ducking powerlines, trying to rescue people.

'You need to tell people this is not a flood: this is a disaster. We're a people who are used to flood. Lismore floods. This is not a flood – this is a catastrophe. We've never experienced anything like this before. This is climate change.'

Outside my window in Adelaide, the sky was blue, and a fresh cup of coffee steamed pleasantly on my desk. It was a stark contrast to the chaos unfolding in Lismore on the other end of the phone. Most people speak to reporters when they have some distance from an event. Sometimes it's hours later. Other times it's a few days or weeks. Higginson was experiencing a tragedy unfolding in real time.

But Higginson, Kate and the people of Lismore were not alone. As the disaster continued to unfold over the next two weeks, an ocean seemed to pour from the sky onto South East Queensland and northern New South Wales. Areas that had burned just two years before flooded, temporarily wiping towns from the map. Thirteen people died in the Northern Rivers area between February and April 2022 as a result of the floods; nine more died during floods elsewhere at that time.

A year later, when I visited Lismore for the anniversary of the flood, I found a disaster still unfolding. Looking around, it was easy to get the impression things were going well. Two in three businesses had since reopened, and considerable work had been done to rebuild. There were still a few boarded-up buildings, but traces of the floods that swept through Lismore had mostly been scrubbed from the streetscape as a point of civic pride.

When my photographer and I crammed into Sue Higginson's idling trayback Toyota LandCruiser for a tour of the area, she said we shouldn't be fooled by what we saw in the town centre.

'On the other hand, you've got the north and south communities, which you can drive around now and still look like an apocalypse hit,' Higginson said. 'And all the evidence says it's going to happen again one day.'

Higginson has a long association with Lismore. Back in the 1990s, she was a frontline forest activist during the era when hippies and environmentalists became a sizable demographic in the small country town. Sometime later, she qualified as an environmental lawyer. Today she and her partner are dryland rice farmers, and she serves as a member for the Greens in the New South Wales upper house.

During the 2022 flood, her family faced a choice: stay or go. She had been paying attention to IPCC reports since the mid-1990s, so the decision to evacuate early was no choice at all.

As we drove over the bridge into North Lismore and turned off the main road, she pointed out the house she'd lived in as a young radical and the site of the old information centre the activists ran. All of the trees around there had been planted in a reclamation project, she said. It all went under in 2022, but the trees had helped break the flow of water so it didn't move so swiftly. Even then, the damage was extreme. Her friends still lived across the road from their old house, she said, and they had endured much. Those who hadn't left before the water rose were forced into their roofs, where they had to wait for help.

The worst-hit houses – the total wrecks – had mostly been pulled apart and hauled away, but traces remained. Get off the main thoroughfares, Higginson said, and it was possible to find them. Hidden between the lush, long grass and the thick tree cover were houses cut with mould and raised homes knocked off the poles that served as their foundations. Richmond River High School was surrounded by rent-a-fence and slated for demolition. It sat right at the lowest-lying part of town, and during the flood the massive structure was totally engulfed.

'It's written off,' Higginson said, idling out front. 'It's never coming back.'

A few streets away, she pointed out cinder block constructions slowly being swallowed by vegetation. There was no glass in the windows, and likely no running water; these places should have been uninhabitable, but cars in the driveways suggested someone was still living there. With the Northern Rivers Reconstruction Corporation dragging its feet on its recovery scheme and the cost of housing rising across New South Wales, most people simply had nowhere else to go, so they went back to the places where they had nearly died.

Over in South Lismore, the 'For Sale' signs were up. Higginson took them as the sign of a broken people. The flood had hit harder in these parts because it was supposed to be safe there. Most residents thought that if their house was above the 1974 flood line, it would all be well.

'These people had no fucking idea a flood this bad could happen,' Higginson said. 'They were people who did everything right. They went to check with council, they built their houses to the right specifications, they raised their homes accordingly.

'They were also the people who were left on their roofs thinking they were going to die.'

Her voice was a mix of anger, hurt and sadness. At points she struggled to hold back tears. When she spoke about these places, she was talking about people she knew. Next she took us to the house her daughter had been living in when the flood came, just a couple of streets away, a 'typical, gorgeous, young-person share house'. Pulling up out front, she pointed out its design features, before adding that the whole thing had been entirely underwater. Luckily, her daughter hadn't been home at the time.

Rumbling back over Leycester Creek, Higginson showed us an empty workspace that was once a Boral facility. Before the flood, the company had stored rusting tanks of petrochemicals in the industrial yard, which it used as a transit hub. During the 2017 flood, the manager had the presence of mind to secure or move anything that might pose a contamination risk. His successor had not been so thoughtful, and in 2022, when the water crashed down through the nearby creek, the bitumen and other chemicals stored in the tanks were discharged. As the floodwaters grew, they were churned into a toxic, muddy cocktail.

A spark in the house across the road had ignited the slick. Photos of the burning house surrounded by water, in the middle of the worst flood on record, became a symbol of the disaster. The house was long gone but, a year on, the woman who owned it was still there, living in a caravan tucked into a corner of her property.

Another house across the road survived, but the bitumen – a solidified form of oil – had blackened the interior walls and contaminated the surrounding soil as the waters withdrew. That house belonged to Marion Conrow, an artist. In the lead-up to the anniversary, she had moved back inside, as the caravan where she lived had no working plumbing. After two weeks, her lungs began to burn. Though Conrow was negotiating compensation with Boral, the company wanted her to sign forms saying they wouldn't be liable for unexpected future health costs. It was a reminder of how the harms caused by oil and gas weren't just the cumulative effect of climate change; often they were local and immediate.

At the time of the flood, 44,334 people were thought to live in Lismore; at the anniversary, no-one was sure how many remained. As the milestone approached, there was talk of exodus. People too traumatised to stick around or those sick of waiting for the government to buy back their homes were selling up cheap, if they hadn't already walked away. Advertisements listed in the front windows of real estate offices hinted at what was happening. One read: 'Opportunity Awaits'.

'This flood-affected diamond-in-the-rough is ready for you to take it to the next level,' it said. 'This classic, high-set hardwood three-bedroom home has incredible potential for a first home or as an amazing investment promising solid rental yield for years to come.'

Investors, property developers and those willing to gamble with their lives on the flood plain to get into an affordable home were swooping in to snap up a bargain. Compared with Sydney, the $225,000 asking price for a home in South Lismore was a steal. Higginson said she couldn't blame the buyers. The frustrating thing, she added, was that it was all so predictable.

From a strictly scientific perspective, the precise role climate change played in the 2022 floods across the Northern Rivers is difficult to determine. A CSIRO report published in November 2022 described climate change as an 'aggravating factor' but did not directly consider its role. To do so, the report said, the agency would need to perform an 'attribution study' using complex modelling and statistical analysis, but as of 2023 the work had not been done. The CSIRO plans to develop a hydrodynamic model for the Northern Rivers area's Richmond River catchment by December 2024 that could be used to 'test any scenario including impact of climate change on future flooding', and as part of that work an assessment *may* be done on the 2022 flood.

In newsroom terms, that means the Lismore flood will be ancient history before the CSIRO is ready to make any further comment. Atmospheric scientists with the US National Center for Atmospheric Research (NCAR) investigated the situation with much more alacrity, publishing the results of their modelling in May 2023. Their work connected the Black Summer bushfires – an event the CSIRO had previously linked to climate change – with the triple-dip La Niña that had resulted in Lismore flooding. In effect, NCAR's researchers said, the 2019–2020 bushfires were so extreme that smoke and particulate kicked into the atmosphere had interacted with the clouds above the Pacific, cooling the ocean's surface and creating the necessary conditions for a La Niña event. The smoke persisted in the atmosphere for seven months, and the feedback loop it set up locked the Pacific into a weather pattern that played out over three years, culminating in the Lismore flood. In that way, climate change could be said to have had a role in the event.

But the NCAR report wasn't enough to make up for the CSIRO's silence. Limited word counts and the rules of objective journalism impose certain constraints on reporters working within the inverted pyramid. The absence of a single, authoritative, quick-reference report from the nation's peak scientific body meant the best they could do was to cite US scientists working for an institution most Australians have never heard of. Neither could they report with any certainty that climate change was responsible for the flooding. The best they could say was that, on the balance of evidence, the American scientists thought climate change was very likely to have made the flooding worse.

And therein lies the trap. When people asked the question 'Did climate change make the floods worse?' the response was never a firm 'yes' or 'no'. The human brain craves certainty, and the ambiguity of any conclusion phrased in terms of risk and probability left space for doubt. As science communicator Dr Lucky Tran has pointed out, when scientists say something is 'very likely' to have made a situation worse, the general public hears the answer to their question as a hard 'no'.

Entire public relations strategies have been built around exploiting the professional demands made of reporters and scientists. Summarising the reality of climate change into one or two paragraphs is not easy, so it ends up being truncated to something like: 'Climatic systems are complex, and it is difficult to attribute any one weather event to climate change.'

I have watched bureaucracies and bad-faith actors run amok in this information gap. When a New South Wales parliamentary committee investigated the state's flood response, the words 'climate change' appeared just once in its report – and only because a Greens MP fought to have it included. Similarly, when Lismore City Council tried to issue a statement thanking the volunteers who helped in the recovery effort, one local councillor, 'Big Rob', led a four-person bloc to have the words 'climate change' scrubbed from the statement so as not to 'politicise' the event. He later told reporters – without a shred of irony – that he did not deny climate change but thought it was a good opportunity to 'stir up lefties'.

'That motion was about thanking people, not being political about climate change,' he said.

2

You Can't Dump Your Rubbish Here

The rain was still falling when word went around that the prime minister was coming to town to tour the aftermath. Kate Stroud and the other Lismore residents were in the process of stripping out all their possessions, ripping down the walls of their homes and heaping their lives into mountains lining the streets. Mattresses, children's toys, metal frames, corrugated iron, utensils, fried television sets, busted cabinets, broken chairs, masses of soggy, wet textiles – the mounds of mud-tainted possessions grew by the hour.

Meanwhile, there had been no true support from state or federal authorities. On the news, it was announced that the Australian Defence Force (ADF) was being dispatched to help, but the only time anyone saw them in action was when a couple of soldiers in clean khaki were dragooned into staging a photo shoot for ADF publicity. As the people of Lismore waited, the systems they relied upon began to fail. Anyone with a car still running couldn't get petrol. No ATMs or EFTPOS machines worked. Finding food was a struggle.

'When you're in a situation where you find a can of tuna in the mud and think that's all you're going to have to eat for a little while, your perspective changes,' Kate says.

To the residents, it felt like the government was holding off declaring a natural disaster, which would have opened up new avenues for support. The prime minister, Scott Morrison, was in isolation, recovering from a Covid-19 infection, and when he finally flew into town, ten days after the flood's peak, the perception was he had been dragged there. When he finally wheeled around to the Lismore Council building, some residents had gathered to greet him in protest. Under the watchful gaze of 40 police officers – a heavy police presence in little Lismore – the protesters chanted slogans like 'The water is rising, there's no more compromising.' Among them was Kate, who held a sign that read, 'Announce a CLIMATE EMERGENCY.'

She had so many things to say to her prime minister, but these boiled down to two questions: *Why won't you help us? And why won't you say this is climate change?* Kate thought she already knew the answers – after all, Morrison was the bloke who'd waved around a hunk of coal in federal parliament. Even so, she thought he should still front up and speak. Had he let them yell at him for ten minutes, he might have earned her respect.

But this time around, there would be no walk down the main street, where flood survivors might refuse to shake Morrison's hand, as they had following the Black Summer bushfires. The prime minister would give them no opportunity to embarrass him in front of the television cameras. Instead, he would bypass them entirely and slip in through the back door. A few hours later, video of a closed-door address was uploaded online. It showed a speech he had given alongside federal emergency management minister Bridget McKenzie. From a room inside the Lismore Council Chambers, Morrison declared the disaster a 'one-in-500-year event' and announced some additional financial support. When it was over, he left, again by the back door.

It was cowardice, Kate felt, and when she was invited onto Network Ten's *The Project* for an interview about the situation, she said as much. The prime minister had announced additional payments of $1000 to the residents, but, as Kate told her interviewers, there was nowhere to spend that money. There were no shops open, no businesses trading. The free market couldn't solve their problems because there was no market. There was nothing to buy. Nothing was working. Had the prime minister listened, he might have been able to offer something useful.

After that interview, Kate received a phone call. A crop of local activists from different interest groups had started talking about how it wasn't good enough for Morrison to duck them. If his house had flooded, they reasoned, he might have seen things differently. Taking this logic a step further, someone suggested they take what was left of their homes to the gates of Kirribilli House, the prime minister's official residence in Sydney. Others had done it after the bushfires – why couldn't they? An election was coming up, and the spectacle might keep Lismore from being dropped in the news cycle. To make it happen, they started asking around for residents who'd stayed in town as the waters rose and were willing to speak up. Everyone had seen Kate's interview, so she was the first to come to mind.

Someone had tipped off the police that they were coming. It was a crisp, clear Monday morning, about two weeks after the torrential rains, when activist Violet CoCo picked up the nine Lismore locals from a Sydney hotel. It was an early start. They left before sunrise to time their action with the morning news.

CoCo's white Toyota Tarago soon joined the small convoy that had driven down from the Northern Rivers. Kate was there too, in the lead car. It was a nervous ride over. People were excited, but they also cried as they followed the truck filled with flood rubbish – a tipper hired from Kennards – towards Kirribilli House. It contained what remained of their lives – homewares, flood-ruined carpet, wrecked furniture, children's soft toys, Christmas trees – and they were about to bring it to the prime minister's doorstep.

About a block away, they discovered police had set up a checkpoint. The night before, protest organisers had sent out a press release under embargo, and a reporter must have checked with law enforcement about what arrangements were being made. Kate's car was stopped by an officer who did not look happy to be awake before dawn. He asked the driver where they were going, and the driver, thinking quickly, said they were 'going to see the sunrise'. Sitting in the back seat, Kate leaned forward to ask what was going on.

'Some climate protest at Kirribilli House or something like that,' the officer said.

'That's a bit annoying,' she answered.

Word filtered through their convoy about the checkpoint, but then, as the cars regrouped nearby, the truck was stopped as it tried to reroute. Undeterred, Dee Mould – an activist who had conducted rescues as part of the Tinnie Army during the floods – climbed atop the truck and began throwing its contents into the street so the others could drag their battered belongings up the road to Kirribilli. As Kate began to haul her things up the road, an officer tried to intervene, telling her to stop. She could see he was 'teetering on the wall between doing his job and battling his compassion', so she asked him a simple question: 'Why?'

'You can't do this,' he said. 'You can't dump your rubbish here.'

'This isn't rubbish,' Kate said. 'This is my home.'

The officer paused. Within seconds, his resistance 'melted' and he turned and walked away.

When they reached the prime minister's house, CoCo stood back to watch. This was the same building Greenpeace had invaded in 1997 to install solar panels on the roof ahead of the Kyoto climate conference. She felt honoured to facilitate but was wary of inserting herself into someone else's struggle – until she was asked to help. Together the Lismore locals piled up the mud-covered remnants of their lives in front of the locked gate and occupied the driveway. Then they posed for photos with a door from one of their houses propped up on a chair in front of them, upon which were written the words: 'Morrison, Your Climate Mega Flood Destroyed Our Homes.'

In the background, police looked on.

Kate and the other Lismore residents would end up on every evening news program in the country. She told one reporter that what had happened in Lismore was the product of climate change, and that she wanted the government to properly fund a transition away from fossil fuels.

'We have basically lost our entire town,' she said. 'We tried to have this conversation with [Morrison] face to face in Lismore, but he slipped through the back door of our council chambers. If our leaders can't come to at least sit at a table with us and chat to us at times of devastation, what are they doing?'

Unbeknown to the group, Morrison had slipped away once again. They had started early, but the prime minister had left even earlier to fly out to Queensland. During a press conference later that day, he would be asked what he thought of the protesters on his doorstep. Rather than respond to anything Kate had said, Morrison told reporters that $1.7 billion had already been committed to flood response and recovery.

'I think the politicisation of natural disasters is very unfortunate,' Morrison said. 'Everyone's just doing the best they can.'

Violet CoCo had been in the middle of planning a series of protests that would shut down Sydney and potentially land her in jail when the phone call came asking if she would escort the Lismore locals to Kirribilli House to dump what remained of their lives on the prime minister's doorstep.

She and a group of other Fireproof Australia climate change activists were living together in a share house, where they had been pulling 13-hour days to plan what they called an 'action block'. These were sustained periods of protest carried out over a week or fortnight to place maximum pressure on authorities to act on a particular cause. The rationale was simple enough: normally politicians did everything they could to ignore a protest, but running multiple actions across a single city several times a day over a period of weeks made the protesters impossible to ignore.

Fireproof Australia is a group that splintered off from Extinction Rebellion in May 2021 following the Black Summer bushfires and was still unknown at the time of the Kirribilli protest. It is often associated with the more militant Blockade Australia, whose members seek to block fossil fuel exports by obstructing physical infrastructure. By contrast, Fireproof Australia is intended to be more accessible to the average person, more focused in its demands, but at the same time more disruptive of people's daily lives. The group still seeks an end to fossil fuel production but it also has three clear, tangible demands, developed through consultation with climate scientists, emergency responders and others. These include a permanent, Australian-based air tanker fleet to fight bushfires, an immediate plan to rehome flood and fire survivors, and the smoke-proofing of kindergartens, schools and aged care facilities to protect children and the elderly.

The Black Summer bushfires were pivotal for CoCo and her fellow activists. They considered oil, gas and coal producers – with their relentless drive to drill more wells and strip-mine for more coal – directly responsible for the death and destruction caused by the fires. These companies had known about climate change and its consequences for longer than most of the activists had been alive, yet they continued to extract fossil fuels and had increased their capacity to do so. The activists' feeling of betrayal only deepened when, after the fire, came the flood.

Everyone at the share house had watched what happened in Lismore. CoCo first learned about the disaster when she walked into the lounge room on the Monday morning to find her best friend, Daisy, in tears. Daisy was a Lismore girl and had pulled up the livestreams on her phone after hearing about the unfolding situation back home. She watched in real time as her friends broadcast the devastation and recorded their efforts to rescue traumatised people from their rooftops. The scale of the disaster was overwhelming.

It was a sobering moment for CoCo and the others. Everything they had been talking about had come to pass.

Again.

As much as Violet CoCo and other activists might point the finger at fossil fuel producers as morally liable for this devastation, actually attributing any sort of legal liability for their activities is another matter entirely. Australian law does not currently recognise anything criminal in helping to generate the carbon dioxide that has contributed to the breakdown of climatic processes, amping up extreme weather events to the point where they increasingly generate a body count. That is not to say litigation is impossible, or that no-one is trying to hold fossil fuel producers to account, but for the most part these lawsuits are civil, not criminal.

Across the world, roughly 2000 climate change lawsuits had been attempted as of 2022. Probably the most famous was the groundbreaking decision by a court in the Netherlands in 2021, when a group of shareholders brought a case against Shell in an effort to force the company to lower its emissions. Despite the odds, they won.

The decision was a turning point for a climate movement that was fast confronting the reality that governments were simply not acting, but it was another battery of cases filed in the US that really started to build momentum. Known as 'They Knew' lawsuits, these litigations specifically charged oil companies and their industry associations with having actively researched the greenhouse effect, concealed the resulting information and then lied about what was actually happening to ensure their future profitability. Of more than a dozen such cases, one of the biggest was *State of Minnesota v American Petroleum Institute*, filed in 2020. In that matter, the US state of Minnesota alleged that the American Petroleum Institute, ExxonMobil, Koch Industries and others had been involved in a 'multi-pronged campaign of deception' that spanned decades. As a result, it was seeking compensation and injunctions.

But it was a separate application, filed under the *Racketeer Influenced and Corrupt Organizations Act* (RICO) in the district of Puerto Rico, that broke new legal ground. Historically, the RICO statute has been used against genuine gangsters like the Gambino crime family and international soccer body FIFA, but *Municipalities of Puerto Rico v Exxon Mobil Corp* was the first time it would be applied to oil, gas and coal producers, after sixteen Puerto Rican municipalities banded together to take them on. Filed in 2022, five years after a deadly hurricane season levelled vast tracts of the island, the suit named ExxonMobil, Shell, Chevron, ConocoPhillips and even Australian miner Rio Tinto among the respondents. The applicants claimed that the municipalities of Puerto Rico, 'believing that the purchase and use of Defendants' carbon-based products were safe and would not endanger the lives or livelihoods' of their citizens, had 'relied upon the Defendants' misrepresentations to their detriment'. As the filing itself described the situation, 'The 2017 Atlantic Hurricane season completely wrecked each Municipality's entire infrastructure, leaving its citizens without electricity for months, stranded from medical care, causing thousands to suffer and die, and rendering them homeless.'

In essence, Puerto Rico's local councils were alleging that the oil, gas and coal companies were gangsters, too.

There is nothing comparable to the RICO statute in Australian law, but the country has nevertheless quietly staked out a position next to the US as the second-most active jurisdiction for climate litigation. According to Melbourne Climate Futures, a multidisciplinary initiative established by the University of Melbourne to contribute to greater action on climate change, there have been 508 climate litigations across Australia as of 2024. Legal academic Rebekkah Markey-Towler, who manages the database, says Australia is unique in that it has both a large body of strategic and non-strategic climate litigation. 'This includes disputes brought pursuant to planning law provisions,' Markey-Towler says. 'Such cases are not necessarily motivated by climate concerns but are relevant to climate change. For example, developers may seek to build homes in areas that are vulnerable to flooding, sea level rise or bushfire risk.'

Every now and then, something really big comes along. One of the most significant Australian climate cases in recent times was *Sharma and others v Minister for the Environment*, in which eight children attempted to force the federal environment minister to consider the rights of future generations when deciding whether to approve a new coalmine. Though they won in the first instance, the decision was overturned on appeal. When the judgment was delivered, the Full Court of the Federal Court of Australia found the minister had no such duty at law, but the judges were divided as to why, all three giving different reasons.

Where the litigants in *Sharma* went after the Australian

government, to date, no Australian lawsuit has taken direct aim at the Australian oil and gas industry for their role in distorting the understanding of climate change. This is partly due to a lack of evidence showing what industry figures knew about the issue and when they knew it. In North America and Europe, it's a very different scenario: committed researchers like Kert Davies at the Center for Climate Integrity and Carroll Muffett at the Center for International Environmental Law have been digging up documents showing what companies like Exxon, Shell and BP knew for the better part of a decade. Their research has gone on to inform several lawsuits seeking to hold the oil industry accountable.

Here in Australia, anyone considering doing the same must negotiate the tricky procedural complications thrown up by Australian law. Anyone even considering filing an application first has to confront several thorny questions. For instance, whom exactly do you sue? Governments for failing to do anything meaningful about climate change? Or corporations for damages? And there are questions of standing, too: do those making the complaint even have a reliable basis from which to make the claim? There are also questions about the proportion of responsibility: Australian companies and operations might be significant global players, but they cannot be considered responsible for all or most of the harm done by the fossil fuel industry. Even if it could maybe be shown that they were responsible for 0.03 per cent of global emissions over their operational lifetime, how might that translate into a penalty? Or, more precisely, what orders might a court be willing to make? It would be one thing to decide a particular company was guilty as sin; it would be another to decide what to do about it. A court could force the company to open its archives, but what formula might they use so their share of global emissions could be translated into a dollar figure for compensation?

Then there is the question of how to actually frame the complaint. Negligence might be an obvious starting point, but Australian legal expert and climate advocate Tim Baxter says such cases are difficult to run, given the 'intentional impediments' that work against success.

'The way I used to describe it to my students when I taught

negligence in tort law was that as you blunder through life you are creating and dissolving a set of relationships around you,' Baxter says. Most of these relationships, however, are not 'special' enough to attract a duty of care. This takes something more, he says. Doctors owe a duty to their patients, teachers to their students, and manufacturers to their direct customers, but there is no obligation created by merely walking past a stranger on a bridge – even if they are about to jump.

'For an individual or organisation to hold governments or corporations liable for their climate harm, there would need to be some kind of special relationship between plaintiff and defendant that may be difficult to prove in the courtroom,' Baxter says.

Making out an application in negligence might be difficult and costly, but it's not impossible – though there are other avenues that promise more success. In Australia, firmer ground may instead be found in corporate and consumer law, where the problems faced by climate litigators have been successfully confronted before – first, in taking on the tobacco industry.

Neil Francey was a young barrister when, in 1987, he first took on Big Tobacco, representing the Australian Federation of Consumer Organisations (AFCO) in a lawsuit against the Tobacco Institute of Australia (TIA). The basic question, Francey explains, was whether the TIA had made misrepresentations in an advertisement about second-hand smoke. 'What happened is they published this advertisement on 1 July 1986 – 'It was titled, "A message from those who do to those who don't", which proactively questioned the evidence of second-hand smoking.' The advertisement included an additional line advising readers how 'there is little evidence, and nothing which proves scientifically, that cigarette smoking causes disease in nonsmokers.'

In publishing the advertisement, the tobacco industry body appeared to misrepresent scientific research on the risks posed by second-hand smoke. At the time, section 52 of the *Trade Practices Act* explicitly stated, 'A corporation shall not, in trade or commerce, engage in conduct that is misleading or deceptive or is likely to mislead or deceive.' Today this provision appears in section 18 of the *Australian Consumer Law* in substantially the same language. Unlike the law of negligence or human rights, the provision's power is in its simplicity: corporations cannot lie when going about their business. An applicant suing an organisation isn't required to prove causation – that is, to show that one thing led to another – or to prove intent. All that matters is that a corporation has made a claim and that the claim is demonstrably false.

What followed was a five-year legal fight that Francey describes as 'war'. As his clients lacked the funds to bring researchers from overseas, he asked eight experts from around Australia to lay out the scientific basis for the damaging effects of second-hand smoke on a person's health. Against him were a team of lawyers from Clayton Utz and overseas, who flew in experts from around the world willing to cast doubt on any relationship between smoking cigarettes and lung cancer.

From the moment the case began, his opponents' strategy was clear: badger Francey's experts in an effort to undermine their credibility. After Francey called his first expert, the tobacco industry lawyers spent two days 'hacking him to pieces' – a spectacle that would be repeated for each of the eight in turn. Francey responded by calling international experts to give evidence.

What Francey didn't know was that the pushback from the tobacco industry was itself part of an organised disinformation campaign. In June 1977, the heads of the major tobacco companies had met for a 'strictly confidential' meeting to discuss an international campaign to challenge scientific evidence linking smoking to cancer. They even gave the campaign a codename: 'Operation Berkshire'. The details would eventually be revealed during the 1990s, when lawsuits in the US forced the industry to make entire troves of documents public.

In the end, the tobacco industry's efforts in Australia failed. In 1991, when a decision was finally delivered in *AFCO v TIA*, Justice Morling found that the tobacco industry had lied.

'The epidemiologists called by the applicant were most impressive witnesses. Their opinions are shared by many other distinguished epidemiologists who took part in the compilation of the major reviews,' Morling said.

'The respondent did not call one witness whose primary expertise was in the field of epidemiology. [...] I have no hesitation in preferring the opinions expressed by the applicant's witnesses.'

The decision established the legitimacy of the science of passive smoking and set in motion a sequence of events that would prove profound. Smoking was progressively banned from public spaces, and six years after the decision, the Tobacco Institute of Australia folded. It's a precedent Francey says could easily be repeated.

'Say a petroleum company publishes an advertisement which in some respects is misleading or deceptive or unconscionable, and it casts doubt on climate change or global warming. That's an idea which is actional *per se* but is up there in the ether,' he says.

'But if a group of low-lying islands in the Torres Strait brings a class action in Federal Court against an emitter for lying, it would be no defence by the company to say that other companies were doing it as well and you can't prove what our contribution is.

'They're making some contribution, though. That's enough.'

In the year Violet CoCo was born, 22.76 billion tonnes of CO_2 were pumped into the atmosphere. It was 1990, the same year that the IPCC released its first report, confirming that human activity was substantially increasing the atmospheric concentration of greenhouse gases – a significant milestone in humanity's efforts to address the existential threat posed by climate change. Back then CoCo was known as Deanna Henskens-Silsbury, and climate change wasn't really a subject she would think about in the first two decades of life. Born to a Liberal-voting, working-class family, she grew up a child of divorce, splitting her time between Newcastle and Sydney. Her dad was an electronic engineer who built weapons systems for fighter jets, and she spent a lot of time wandering an empty air base she would later describe as a ghost town.

As a teenager she discovered the theatre, and to pay for drama classes she worked fast-food and retail jobs. After high school, she briefly took a job at the bank where her mother worked, and voted Liberal in her first election, like the rest of her family. Later, she landed a job working for a Murdoch-owned advertising company, selling search engine optimisation services to small businesses. It was a job she was 'ridiculously good at', and the bonuses meant she was quickly earning six figures, but she could also 'see the toxicity' in what she was doing and soon quit to study philosophy at university.

CoCo didn't finish her degree and instead took time off to travel Australia. After a year of living in a van, she landed in Melbourne, where on the first morning she was invited to attend a rally in town – a counterdemonstration against a planned march by a group of Neo-Nazis. She hadn't known there were still people around who called themselves Nazis. She thought the world had dealt with that seven decades ago.

'Hell, yes – let's go!' she said.

The skirmishes with police and white supremacists in the streets of Melbourne were CoCo's first real engagement with politics and protest. It was the beginning of a transformative period of her life. As she began to explore this world, she shed the name her father had chosen and called herself 'Violet'. People had been calling her 'CoCo' since she was 18, because of her love of chocolate, and so she became Violet CoCo.

Her first forays into environmental activism were tame by her later standards. She'd always cared about animal rights and the environment, and in her late teens had made regular donations to groups like the Australian Conservation Foundation. As an adult, she had signed petitions – including the biggest e-petition ever delivered to federal parliament at the time, signed by 404,538 Australians, calling on the government to declare a climate emergency. She had also helped organise rallies; in February 2020, CoCo was one of the organisers of the People's Climate Assembly protest in which 5000 people encircled federal parliament. Yet all these efforts were ignored by everyone who mattered.

Around the same period, hundreds of other protests, big and small, were taking place across the country in response to incursions by fossil fuel producers. Up in Queensland, protesters opposed to Adani's coal operation squared off with law enforcement. Indigenous groups organised to fight new gas developments on their land. Farmers opposed to fracking worked together under the banner Lock the Gate. Among the biggest protest groups were the School Strikers – some 30,000 children and their parents who marched to demand climate action. The marches made international headlines – at one point, they brought 300,000 people into the streets nationwide – but, in a speech delivered on the floor of parliament, Prime Minister Scott Morrison responded by telling those children to get back into the classroom.

'Each day I send my kids to school, and I know other members' kids should also go to school, but we do not support our schools being turned into parliaments,' Morrison said. 'What we want is more learning in schools and less activism in schools.'

With Morrison in the headlines and a mass fish kill in the Menindee Lakes making national news, CoCo felt a growing sense that 'the earth seemed like it was dying'. It was around this time that she heard about a group calling itself 'Extinction Rebellion' (XR) and attended her first meeting. There she listened to other women talk about the existential threat posed by the climate crisis. She felt moved by their call for nonviolent direct action to stop it – and she immediately volunteered to help.

Her initial contribution was to recruit others, but when she looked at what was happening in the UK, she both felt inspired and disheartened. It was incredible to watch people shut down the City of London by blocking its five bridges, but here in Australia, XR had never held an action transgressive enough to risk arrest. The group's Australian membership was tiny, and CoCo suffered from impostor syndrome as they had not actually done anything. For all the talk about the duty to protest and the necessity of risking everything, she was acutely aware she had no skin in the game. She herself had never even been arrested.

She would later look back on her first protest with XR as a 'disaster'. Twenty first-time rebels had glued themselves to the windows outside the Independent Planning Commission offices in Melbourne to object to new fossil fuel approvals. Using glue made it harder for the police to haul them away and bought the activists more time to cause a scene. Acetone would later become a standard-issue part of police kit.

But on this occasion, as the protestors weren't technically blocking anything, the police were content to watch politely from a distance. With nothing actually going on, the television reporters in attendance threatened to pack away their cameras and leave the activists to it. Thinking quickly, CoCo dispatched a member of the protestors' support group to get spray paint, so she could paint 'Climate Emergency' on the glass. It arrived 40 minutes later. Soon after, she experienced her first arrest.

That moment was the real start of her political education. Looking back, CoCo would later describe herself as having been 'politically illiterate' until she joined the climate movement. She had thought the Adani coalmine – a massive 279-square-kilometre open-cut coalmine in Queensland, capable of producing 10 million tonnes of coal a year – was the only mine in Australia. There were in fact over 350 operating mines in Australia as of 2020, 94 of them coalmines. The more she learned, the more questions she had. Why was it that neither Labor nor the Coalition seemed to take climate change seriously, when the dangers were so clear and so obvious? Why was it that climate scientists couldn't get meetings with political leaders, while the fossil fuel companies seemed to get whatever they wanted? How was it that South Australian oil company Santos could forge ahead with its plan to develop the Narrabri gas field, drilling more than 700 gas wells in the Pilliga Forest?

The more she learned, the more she found that others also lacked the same basic knowledge she did. It was an observation that hit home during another protest in Melbourne, around the time of her tenth arrest. CoCo was leading a small group of inexperienced activists in a protest outside APPEA's local office. The plan was to block traffic on the road out front by dropping three individual sections of fence, arranging them in a triangle and then having the activists lock themselves to the corners by the neck so they couldn't be easily moved on. It was all supposed to be relatively straightforward, but then someone asked something unexpected: why APPEA?

CoCo was initially taken aback. How could they not know? APPEA were the villains, she tried to explain. They were the visible manifestation of the oil industry's influence; the political arm of a producer group that allowed individual companies to stay out of the ring when it came time to get into the mud. Most of all, APPEA was effective – a reputation sometimes spoken of with awe. They were a band of slick, super-sophisticated lobbyists, with all the money and all the power, and they always seemed to be one step ahead of the good guys.

She knew APPEA sat at the point of intersection between fossil fuel producers and government. They lobbied, but they weren't just lobbyists. They were industry, but they didn't produce. They palled around with politicians but weren't elected representatives. When CoCo tried to explain all this, people didn't really get it.

'The people in the climate movement don't realise how illiterate on the issue the general public is,' she says. 'One of the most annoying things about being in this work is that you find out your planet's dying and suddenly you have to be an expert in all the ways the corruption is happening, and all the bodies that are doing it, and all the politicians that are doing it, and there's just so much to unpack about how we've got ourselves in such a difficult situation.

'And then you have to be able to express that to people in less than thirty seconds.'

3

A Mob of Bastards

The first thing to understand about the early Australian oil industry is that it was late to the party. From the moment Winston Churchill announced the conversion of the British naval fleet from coal in 1911, oil was destined to take over our lives, and by 1931, eight wells a day were being drilled in east Texas alone. In Australia, there was nothing. Oil wouldn't be found until 1953, and even then it still took another decade to really get moving.

The second thing to know about the Australian oil industry is that the US companies whose financial might and technical expertise underwrote its development did so knowing that burning fossil fuels threatened the chemical composition of the atmosphere. You can pick your date for when they became aware of the risks – there are several. What matters is that, by the time Australian petroleum production started up, the global industry knew there was a problem, understood the risk and went ahead anyway.

Take, for example, the flurry of activity that took place from 1958 onwards. The earliest work by industry on climate change evolved out of research undertaken into the origin of smog above Los Angeles during the 1940s and early 1950s. As the public began to connect the act of burning petroleum to pollution and a changing environment, the American Petroleum Institute (API) quickly caught onto the implications. By 1958, its internal Smoke and Fumes committee had engaged Truesdail Laboratories to run 'collection and analysis of gaseous carbon compounds in the atmosphere to determine the amount of carbon of fossil fuel origin'.

But it was another meeting at Columbia University in November 1959 which provided one of the earliest known warnings about what burning oil, gas and coal actually meant. The API, which represented the entire US oil industry, and the Columbia Graduate School of Business, brought together 300 government officials, economists, historians, scientists and industry executives for a symposium in New York, titled 'Energy and Man'. The event was convened to celebrate a century of oil production in the US. Among the presenters was physicist Edward Teller, who had helped develop the nuclear bomb and later denounced J Robert Oppenheimer as a communist.

For all his faults, Teller was at least honest about the obvious. Looking out over the horizon and into the future, he told his audience that the problem of building an economic system around oil was that it ran out. As demand grew, the cheapest and easiest fuels were extracted first, and the cost of finding new supplies grew, driving up the price of fuel over time.

'But I would first like to mention another reason why we probably have to look for additional fuel supplies,' he went on. 'And this, strangely, is the question of contaminating the atmosphere.'

Teller mentioned ongoing work by the Scripps Institution of Oceanography in La Jolla, California. This was presumably a reference to work by scientist Charles Keeling, who was taking early measurements of atmospheric carbon dioxide – and who received seed funding for his initial work by a coalition of oil and car manufacturing companies – but had not yet published his findings. Teller then outlined the basic physical reaction produced by burning oil, gas and coal for fuel, and the 'strange property' of carbon dioxide: 'It transmits visible light, but it absorbs the infrared radiation which is emitted from the earth. Its presence in the atmosphere causes a greenhouse effect. It has been calculated that a temperature rise corresponding to a 10 per cent increase in carbon dioxide will be sufficient to melt the ice cap and submerge New York.

'All the coastal cities would be covered, and since a considerable percentage of the human race lives in coastal regions, I think that this chemical contamination is more serious than most people tend to believe.'

During questions, the dean of the Columbia School of Business, Courtney C Brown, asked Teller to 'summarise briefly the danger from increased carbon dioxide content in the atmosphere in this century'. Teller responded by taking a stab at it with a quick, back-of-the-envelope calculation. 'At present,' he said, 'the carbon dioxide in the atmosphere has risen by 2 per cent over normal. By 1970, it will be perhaps 4 per cent, by 1980, 8 per cent, by 1990, 16 per cent, if we keep on with our exponential rise in the use of purely conventional fuels. By that time, there will be a serious additional impediment for the radiation leaving the earth. Our planet will get a little warmer. It is hard to say whether it will be 2 degrees Fahrenheit, or only 1, or 5.

'But when the temperature does rise by a few degrees over the whole globe, there is a possibility that the ice caps will start melting and the level of the oceans will begin to rise. Well, I don't know whether they will cover the Empire State Building or not, but anyone can calculate it by looking at the map and noting that the ice caps over Greenland and over Antarctica are perhaps five thousand feet thick.'

These amounted to throwaway lines in a talk mostly spent advocating for nuclear power as the fuel of the future. Along the way, Teller suggested that nuclear weapons might be used for 'earth-moving jobs', such as the construction of harbours and canals, on the basis that 'a nuclear explosion is cheap and big'. It was an idea that would later captivate Australian mining magnate Lang Hancock and his daughter, Gina Rinehart.

It is unknown what the reaction was among Teller's audience at the symposium, but the American oil industry was paying keen attention. In the ensuing years, the API began to direct scientific guns for hire – particularly those with the Stanford Research Institute (SRI) – to privately look into all this guff about the greenhouse effect. Unfortunately for the oil executives, the US government had conducted its own investigations and in 1965 the issue went public. It's unclear when, but after the research into city smog and pollution began to raise concerns about the effect of carbon dioxide, US President Lyndon B Johnson had instructed his own group of scientific advisers to look into the matter. When this committee reported back, they had clearly identified a 'measurable' increase in CO_2 and carbon monoxide levels in the atmosphere, and they attributed the increase directly to burning oil and coal.

Three days after this report was delivered, API president Frank Ikard addressed his membership at the organisation's 45th Annual General Meeting to reassure them the industry had the situation well in hand. 'One of the most important predictions of the report,' he said, 'is that carbon dioxide is being added to the earth's atmosphere by the burning of coal, oil and natural gas at such a rate that by the year 2000 the heat balance will be so modified as possibly to cause marked changes in climate.'

To soothe his audience, Ikard explained that API had been sponsoring research and publishing its findings for more than 10 years, 'starting long before the national spotlight was thrown on this problem'. He also made it clear that the oil companies had resolved to address the issue.

'Our industry is best understood, and its needs are more readily appreciated, when it can strike a single note,' he said. 'Our industry thrives today because it has identified those areas that could help, or impede, its progress; and when it has responded, it has done so with a clear voice, easily understood.'

Ikard's speech would be written down and published for distribution, but it soon faded from memory. Meanwhile, API turned to the SRI scientists for help. SRI was affiliated with Stanford University but was not a typical research department. As a contracting lab, its purpose, in its own words, was to 'serve industry' – not dissimilar to the CSIRO. Earlier in the decade, the oil industry had turned to SRI's research teams when it wanted to poke holes in the idea that car exhaust caused city smog. Now they wanted to find out what was going with the greenhouse effect.

SRI scientists Elmer Robinson and RC Robbins conducted early atmospheric research for API and the wider US oil industry. Their final report, delivered in 1968, was damning. In its pages, the researchers told the oil executives that carbon dioxide was the only air pollutant 'proven to be of global importance to man's environment on the basis of a long period of scientific investigation'. The scientists were also clear about the cause of increasing CO_2 levels, saying, 'Although there are other possible sources for the additional CO_2 now being observed in the atmosphere, none seems to fit the presently observed situation as well as the fossil fuel emanation theory.'

And they went further, even outlining the potential impacts: 'If the earth's temperature increases significantly, a number of events might be expected to occur, including the melting of the Antarctic ice cap, a rise in sea levels, warming of the oceans and an increase in photosynthesis.'

But they were not done. The researchers also delivered a stunning conclusion: 'Man is now engaged in a vast geophysical experiment with his environment, the earth,' adding, 'Significant temperature changes are almost certain to occur by the year 2000 and these could bring about climate change.

'There seems to be no doubt that the potential damage to our environment could be severe,' they said.

The API didn't tell the public what it had learned. Instead, it asked the researchers to quickly produce a second, 'supplemental' report that was more cautious in its conclusions. The researchers delivered this second, softer report to the World Petroleum Congress in 1971. Over the next ten years, their work would be cited again and again by in-house teams keeping an eye on the issue for their employers. Eventually, it would be identified by the Center for International Environmental Law (CIEL) as showing that the US oil industry knew about the risk and then went on to lie about it. According to the CIEL's director, Carroll Muffet, the 1968 SRI report represented a clear delineation between 'before' and 'after'. This was not evidence from a single company, but from an industry association. Big oil in the US was implicated to its core.

Decades later, the global oil industry – including the Australian oil industry – would strike a somewhat different tone. Faced with a growing realisation among the public of the threat posed by burning fossil fuels, the industry would run a global disinformation campaign to pollute the public discourse, claiming science didn't know enough to justify action. This was a lie. They had known the risks for half a century. They were just responding to something far more powerful than the laws of physics: the iron will of the stock market.

How an oil company is valued has changed over time, but it's usually about measuring risk. Finding oil is one thing; getting it out of the ground is another. Working out what any given project might be worth means figuring out what's there, eyeing off the oil price, guessing at the costs involved – labour, transport, tax, technology, finding a buyer – and then factoring in the unknown to work out the probability of success. If a project worth $300 million on paper is thought to have a 15 per cent chance of ever actually happening, it's considered to be worth $45 million. Try as they might, however, there is always doubt. Every time an executive pulls the trigger on a new development, there is a degree to which they are gambling – but this uncertainty has never stopped them. They have pulled that trigger again and again.

Once up and running, an oil company simply has to drill. It is a non-negotiable financial imperative. The more oil or gas a company extracts, the faster its fields are depleted and the more urgent the need to find replacements. The problem is existential. Put simply, no more oil means no more oil company. As new fields are found, they are referred to as 'booked assets' or 'equity oil' – resources that have been identified, 'proved up' and are technically owned by the company, which can be moved into production when the need arises. Booked assets are another factor in assessments about the future profitability of a company – and its share price.

The result is a relentless pressure to grow – which is why, by the middle of last century, the largest oil companies in the world began to cast their gaze overseas, to faraway places like Australia, where they would embark upon a massive expansion of their operations, despite what they already knew about the dangers posed by burning fossil fuels.

Australia was originally considered too old for oil. According to the best geological thinking of the time, the sprawling, ancient continent didn't have the right rocks. Colonial-era water wells were periodically contaminated by oily residue, and a bore sunk out near Fitzroy Crossing in Western Australia was found polluted with crude. Oil was even found at Lake Bunga in 1924 near Lakes Entrance in the Gippsland Basin but the source was determined to be coming from offshore. Despite this, the consensus was that the layers of rock in which oil formed had been eroded away over hundreds of millions of years and the sediments were simply too thin for it to form.

A total of 22 exploration wells were drilled in Australia between 1915 and 1953 before this thinking was overturned. World War II had left Australia acutely aware of its own vulnerability, and the Chifley Labor government wanted to encourage 'energy independence' through the creation of a domestic oil industry as part of a broader industrialisation drive. The idea of a national, publicly owned oil company entered the political imagination then – the government even agreed to buy a drill rig for its in-house geologists to work – but it would never be realised. In 1949, when Liberal leader Robert Menzies became prime minister, he immediately killed the idea, despite having supported it in opposition. Doing so guaranteed Australia's oil wealth would be managed by private hands.

Into this world strode William Gaston Walkley, in a powder-blue suit and matching tie. An entrepreneur from New Zealand, Walkley was a heavy-set man with the droopy expression of a basset hound. He had founded the Australian Motorists Petrol Company (Ampol) in 1936 but decided early on it wasn't enough just to run the pumps; he wanted the field and the refinery as well. In pursuit of this goal, he roped in the California Texas Oil Company, known as Caltex – a joint venture of Standard Oil of San Francisco (Chevron) and the Texas Company (Texaco) in the Asia-Pacific region – to form a new joint venture in 1952.

If they wanted oil, the only question was where to look. Mulling it over for a year, they eventually settled on Rough Range in the Exmouth Gulf along the mid-north coast of Western Australia. Walkley was so sure of himself that local reporters in Carnarvon bought him a red ten-gallon hat and made him promise to wear it down Sydney's main drag when he found oil.

The very first shaft sunk struck oil at a depth of 1100 metres – but no-one noticed at first.

It was 1 pm on 1 November 1953 when it happened, and 38° Celsius in the open air. Earlier in the morning they had run a test, and it had taken the roughnecks working the rig about an hour to raise, disconnect and stack each 30-metre section of pipe. It was heavy, time-consuming work, and in the afternoon the men stripped back to stay cool in the hot sun. It was only when they were done that they found the floor of the rig was awash with a hot, waxy, kerosene-smelling green-brown oil.

Their find made geological history, and William Walkley would go down a legend. Standard Oil withheld the news for a month before it finally published the result from its San Francisco office. When it went public, the word 'bonanza' was splashed across newspapers' front pages all over Australia and Walkley wore his red ten-gallon hat down Pitt Street in Sydney, stopping traffic as he went. The flagging Australian oil industry was suddenly revived as the share price of every start-up explorer shot up. A whole new generation of geologists 'got their ticket' when they joined the industry in the rush. The frenzy was such that no-one paid much attention when Caltex quietly reported that eight follow-up wells drilled over the next year had failed to find anything at all. Rough Range-1 itself would briefly produce 550 barrels a day before it ran dry in May 1955.

Prior to the find at Rough Range-1, most Australian oil explorers had struggled to survive. Coasting off the hype, the industry found itself with a second chance at life, but most would still struggle. Their counterparts in the US and Europe had grown into titans over the course of the twentieth century, but the Australian 'oilmen', as they called themselves, had trouble keeping their companies afloat.

★

From the beginning, the basic plan of Australia's wannabe oil barons was simple enough. They were going to be oilmen, they were going to get very, very rich, and they were going to make sure the government made it as easy as possible.

The Australian Petroleum Exploration Association (APEA) began life in 1959, long before anyone had actually managed to develop a commercial oilfield. APEA's entire purpose was to help the nascent oil producers achieve their political ambitions. Although others had floated the idea of an industry association, its genesis is credited to John Fuller. Fuller was an ex-vaudeville performer with a pencil moustache and a 'forceful personality'. He had served during World War II and was captured in the fall of Singapore, where he learned about the oil industry from fellow prisoners during three-and-a-half years in captivity. Once free, he switched degrees from medicine to economics and founded his own oil company, Planet Oil, after graduating.

Looking overseas to the US, Fuller eyed off organisations like the American Petroleum Institute as an example that he and his Australian colleagues could follow. From magazine clippings, he learned that the API was to hold an oil industry conference in Odessa, Texas, in 1958. A tiny town in the desert at the western edge of the state, Odessa was known for two things: oil and cattle. With 300 days of sunshine a year, the city seemed to breathe in with the boom and out with the bust. The year Fuller arrived, a massive petrochemical complex opened to make synthetic rubber, and the city would soon become a key hub in the oil industry supply chain. He paid his own way to the conference, not to browse the exhibitions and sit through tedious technical presentations about the latest drilling techniques, but for the chance to meet industry figures on the sidelines. Over the next three months, he stayed on, travelling the US to meet oil executives and collect industry association charters on which to model Australia's own.

The hot issue in those early days was subsidies. What the Australian industry wanted most of all was government money to help boost exploration, and APEA was initially founded to campaign on this issue. The Australian government had already passed the *Petroleum*

Search Subsidy Act 1957, which set aside £500,000 a year – roughly $18.7 million in 2024 dollars – to co-fund drilling operations. Under the Act, the government would cover half the cost of stratigraphic drilling – drilling done to investigate conditions below ground and identify source rocks for oil – to encourage companies to drill deeper. All information obtained through these investigations would be made publicly available, which would help the smaller players better target their operations.

But that wasn't enough for the wannabe oilmen. The scheme only applied to one type of drilling – and worse, it favoured the big players over the wildcat drillers, who considered themselves plucky individualists whose work was critical to the national interest. Robert Menzies listened to their complaints, and in 1958 expanded the scheme, agreeing to cover two thirds of the cost. For good measure, Menzies even bumped the funding up to £1 million a year – roughly $37 million in 2024 dollars – and changed the program's parameters to include more operators. It still wasn't enough, and the oil companies responded by plotting rebellion.

Fuller had done the early legwork, but the industry's call to arms came in the form of an open letter written by Santos geologist Reg Sprigg, published in the *Australasian Oil & Gas Journal* in June 1959. In his letter, Sprigg laid out his goals for a new industry association that would 'bring together all individuals and groups engaged in, or actively interested in oil exploration' and 'formulate broad policy for Australian participation in all phases of exploration and operation'. Sprigg wanted this organisation to 'maintain the highest ethics in the industry' but also made clear its primary function would be to 'act as a central clearing house and dissemination centre for basic information and propaganda' and to 'educate public opinion by articles and press reports'.

Four months later, a meeting of the industry's heavy hitters was convened in the offices of Planet Oil to work out the mechanics of the new organisation. It needed a leader, and the figurehead they turned to was Reg Sprigg. It was Sprigg's work that had helped overturn the idea that Australia was too old for oil, and he was well respected in the industry. Sprigg, however, had no real interest in politics and had only ever wanted to be a geologist. Reluctantly, he was appointed to serve as founding chairman. Sixteen industry representatives met again two weeks later in a lawyer's office on Hunter Street in Sydney to formalise the arrangement and officially create APEA.

The organisation was an exercise in business unionism. Its immediate benefit was that it created distance between its member companies and government. Senior industry figures could sit on the board or make up its executive, with the organisation functioning as a proxy through which they could act, while keeping their own hands clear of the knife. Each company had its own relationship with state and federal government, but so did APEA's board and executive. Between them, they could share intel, resources, expertise and strategies and coordinate their efforts where it was in their mutual interest to do so. Factions would form within the association, but in public everyone made sure to play nice beneath the APEA banner. If they ever turned on each other, they would be cutting each other's throats.

Membership of the organisation offered a cure for impostor syndrome. Though they hadn't actually found any oil, joining APEA showed they were serious. APEA fostered a certain culture among its members: simply being an oilman was an identity itself, one that afforded status and respect.

In later years, the association actively recruited from inside government, but in the early days that would have been unthinkable. The oil explorers cast themselves as rugged individualists and anti-authoritarians who only cared about finding the next barrel. It made for a noisy community of competing factions, each with their own ideas about how things should be run. Often the main point of contention was which posture to take when dealing with government – assertive, belligerent or cooperative. Some pushed for amiable collaboration; others wanted to take no prisoners. The association would wrestle over this particular question for decades. Often the fissure was between the smaller players and the bigger multinationals who threatened to swallow them up.

APEA would be the first of the two significant oil and gas industry associations in Australia, though it would later be considered a junior partner to the Australian Institute of Petroleum (AIP). Founded in 1976, AIP absorbed three other industry organisations: the Petroleum Information Bureau, the Petroleum Marketing Engineers' Advisory Committee and, crucially, the Petroleum Industry Environmental Conservation Executive (PIECE) – an industry group that, in preceding years, had held seminars across the country to address growing public concern about environmental issues. Where APPEA was decidedly domestic in outlook, AIP was to represent the Australian petroleum industry 'at the international level', even as it maintained 'close liaison' with state and federal governments. On start-up, the institute's founders took advice from their counterparts at the American Petroleum Institute in the US, and the Institute of Petroleum in the UK 'on a variety of matters'. It was, however, the assimilation of PIECE that meant, from the very beginning, AIP was uniquely focused on monitoring environmental issues and neutralising public concern about potential harms.

Until the formation of AIP, APEA was the only show in town. It would take them decades to learn how to properly wield influence – though not for lack of trying. From the beginning, some within the APEA wanted it to pursue a strategy of state capture. Minutes from an association council meeting dated 24 February 1960 record one founding member, Eric Avery – a Cambridge-educated veteran of World War II and personal friend of both US oil baron John D Rockefeller and Chinese nationalist leader Chiang Kai-shek, two of the most influential men of the century – arguing for the organisation to take a more assertive role.

'There is a need to set up routine machinery where consultations occur between APEA and the Government's people [on] matters for initiation by the latter, whereby they have an avenue for coming to seek our view,' he said.

It was an idea that would endure over the years, with other members floating similar proposals. One suggestion was that Australian oil companies should take inspiration from their counterparts in Canada, who had struck an arrangement with government. There,

company executives looking to exploit the tar sands – a vast source
of bitumen that required heavy processing to be turned into crude
oil – would hold personal, off-the-record talks with government
officials. Nothing was written down at these meetings. The private
nature of the conversations meant that views could be expressed that
didn't fit the company or party line. Those involved would then go
their separate ways with a renewed understanding, and policy would
be shaped accordingly. It was, however, a strategy that couldn't be
replicated in Australia at that time; according to APEA, Australian
powerbrokers were 'too emotional'.

But the bigger problem was that the oilmen lacked credibility. It
was only in 1960 that the first commercial quantities of gas would be
extracted from a well 32 kilometres outside Roma, Queensland – the
same region where gas had been found when digging bores for
drinking water. The following year, the first commercial-scale oil
production would begin at the Moonie oilfield, 32 kilometres south
of Cabawin, Queensland, at a rate of 1000 barrels a day.

With these finds, the industry started to get traction, and APEA
went to work. From its foundation, the association had been built
around three committees that made clear its priorities: technological,
government relations and public relations. The technological
committee was chaired by Bill Siller of Lucky Strike Drilling,
another founding member of the association. The government
relations committee was headed by Bruce Graham – who, in a clear
conflict of interest, served simultaneously as both a director of oil
company Papuan Apinaipi and Liberal Party MP for North Sydney,
and used his public position to advance his private interests. But it
was the public relations committee, headed by Alan Prince, another
Papuan Apinaipi board member, that initially had the greatest status
within the association.

From the start, good press was considered vital to good business.
Cultivating the media was treated as critical to securing what
might today be described as the 'social licence to operate'. For oil
exploration to take place, the public needed to accept the presence
of the company in their communities. It also didn't hurt if the public
were active supporters.

And APEA already had good practice at getting their views across in the media. John Fuller regularly contributed to the *Australian Financial Review*, and Prince bought the *Australasian Oil & Gas Journal* to publish industry news on APPEA's behalf. Its brightly coloured pages were filled with thinkpieces, ads and industry gossip, written in the clipped tone of a 1950s radio presenter. Having secured channels through which to present their views to the business community and decision-makers, the next step was drumming up support. Their chosen method was to hold a national conference that would bring together engineers, government geologists, corporate leaders and Australia's political leadership all under one roof.

No-one was particularly interested at first. The Australasian Institute of Mining and Metallurgy passed on a chance to participate in the 1961 conference, and the Geological Society of Australia declined an invitation for its geologists to present their technical papers at the event. Despite these setbacks, APEA booked a room at the Hotel Australia on Collins Street, Melbourne, and planned a publicity campaign that pitched the upcoming meeting as a 'gathering of scientists'. The choice of venue spoke to their ambition. The 12-storey building was the most prestigious hotel in the city and a regular haunt of the Australian elite. Victorian premier Henry Bolte received – and eagerly accepted – an invitation to open the event. Stepping to the dais in the hotel's conference room, Bolte addressed an audience of 70 men in full suits with neatly parted hair and encouraged them – as men of industry and science – to continue their noble work for the good of the nation.

There were formats for running these sorts of events, but no-one involved with APEA at the time really had any clue how to throw an industry party. Four technical sessions were delivered over two days and would have been about as thrilling as a football club AGM or church seminar – though that wasn't really the point. The real money was in networking. When it was over, APEA's membership had grown by 25 companies, and its councillors considered it a resounding success. They had just one regret: government geophysicists from the Bureau of Mineral Resources, Geology and Geophysics had been forbidden by their department head back in Canberra to attend, for fear they'd grow confused about their loyalties.

The industry then took its first real crack at blurring the lines that separated them from government. The oilmen had taken the absence of the government geologists personally. They wanted to be mates, and they couldn't understand why government didn't want to be mates, too. To get around this separation of church and state, the industry resorted to sleaze. As committee head Bill Siller would later recall in an interview, liquor was his weapon of choice.

'The ban was only specific to the conference for technical sessions. It said nothing about social functions,' Siller said. 'I went to them and said: "What about if you come for a drink with Bill Siller? The bosses can't object to that." I received many acceptances, and all of them had a wonderful time mingling with other delegates at the conference cocktails and dinner.'

The next year APEA resolved to address the situation more directly. In an 'emotive' pamphlet that denounced the government for failing to give oil explorers enough support, the industry proposed a four-year plan for its future. Knowing a copy would find its way back to government eventually, APEA organised a dinner in Canberra soon after, to which 'a carefully selected number of parliamentarians were invited'. Among the invitees was Senator William 'Bill' Spooner, then minister for national development in the Menzies government, who was given a chance to make an address. After a few pleasantries, he got stuck in.

'You're a mob of bastards,' he said.

Spooner told the industry that the 'gloves were off now' and those in the room 'should expect no favours from him'. These proved hollow words, however, as Spooner went on to cave entirely to their demands. Not only did he promise to consult more closely with industry in the future, but he lifted the ban on government geologists mingling with their counterparts at the oil companies. It was a total victory. Everyone seemed to fall into line. Australia was a vast landscape and needed oil to be independent. Our oilmen might have been bastards, but they were *our* bastards.

4

We're Not Going to Apologise

It was somewhere thousands of feet above Los Angeles, his plane helplessly circling above the sprawling metropolis, that Liberal Tasmanian senator Denham Henty first had the idea to investigate the issue of air pollution. He had been in the US, on a tour through Fort Worth and Cape Kennedy, and was making his return when it happened. As the aircraft approached, word came that the smog collecting above the city was too thick for them to land.

When he finally put two feet on the ground, Henty was appalled at what he saw. Caltech scientist Arie Haagen-Smit had made the first connection between the two million cars on the city's roads and all that smog that hung in the air in 1948. The petroleum industry fought him over the claim, denying there was any relationship at all, but Haagen-Smit's work was persuasive enough that it started to mobilise action. Still, actually passing the regulations needed to control the problem took time. Even as late as 1958, people wept on LA's streets due to eye irritations, and some resorted to gasmasks to move about outside.

'Don't let this sort of thing happen in your own place,' people told Henty. 'Don't let it happen in your own country.'

Steaming back home, Henty wanted to address the issue before it became a problem in Australia. He decided to muster the collective resources of the senate by creating a committee system that put senators to work investigating practical problems. The first would be the select committees on air pollution and water pollution.

George Howard Branson, Liberal senator from Western Australia, would be chosen to head up the eight-person Air Pollution committee. Over the course of 1968 and 1969, the members toured industrial sites and took evidence from experts at companies like General Motors Holden, Shell, Amoco, Esso, BHP, Alcoa and the corporate predecessor to Rio Tinto. It was on 6 February 1969, at a hearing in Hobart, that they received a startling warning from University of Tasmania professor Harry Bloom.

Under the watchful gaze of the committee, Bloom began his address with a declaration: the threat posed by air pollution generation was 'much more serious than is normally accepted', and especially so the risk posed by carbon dioxide. He was particularly concerned that no-one seemed to be paying attention to 'one very important oxidation product of carbon, and that is carbon dioxide'.

He told the committee: 'Carbon dioxide build-up in the world has been calculated to be such as to be able to produce serious changes, not only in climatic conditions but also in health conditions all over the world in not too many years, say fifty to a hundred years. I think the whole situation is one which needs very desperate and immediate action. I think we have to know what is at present in the atmosphere, and one ought to do something about it.'

The problem was well documented, Bloom later explained, as he had 'seen some very highly scientific studies of this matter'. Unfortunately, he never named the studies he was referring to in his evidence – a frustrating omission, like so many others in the official record of early discussions of climate change, making it difficult to gauge now exactly what was known when.

But there is no doubt that Bloom found what he had read convincing. During questioning, the professor was asked to identify which pollution problem he thought was of most importance, giving him the opportunity to drive his point home:

'If there is a serious and permanent change of climatic conditions in a state or country, or in the world,' he said, 'it could conceivably be impossible to do something about it. If carbon dioxide built up to such an extent in the earth's atmosphere as to trap radiation from the sun and cause climatic conditions to change all over the world, perhaps heating the whole world and melting the ice caps, nothing could be done about it at that stage. At this stage, when we recognise the problem exists, we ought to do something about it before it becomes too late.'

The minutes of the exchange don't record the passion with which Bloom delivered these words, but if he thought he would not be believed, Branson was quick to reassure him.

'You are among friends,' Branson said. 'That is why we are here. The Commonwealth thought this, too.'

Democratic Labor Party senator John Little suggested then that the good professor might be overreacting. 'As a layman,' Little said, it seemed to him 'a rather extreme assumption' that continuing to add carbon dioxide to the atmosphere might result in catastrophic climate change. There had to be some sort of 'compensatory factor', the senator insisted. For example, he said, a hotter planet would generate heat, leading to more evaporation and more rainfall, which might cancel out the warming effect on the poles.

Little was right in a sense: higher humidity on a warmer planet would lead to higher evaporation rates and more torrential rain. But his suggestion that more rain would cancel out the risk of melting ice at the poles was crackpot stuff.

With a nudge from Branson, Bloom answered this 'gotcha', explaining that it was nonsense. Though the professor said he was no pessimist – when it came to predictions about cities being swamped by rising sea levels, he did not 'swallow all of the stories in their grimmest detail' – he was adamant that he was correct.

'There have been analyses of carbon dioxide in the atmosphere over a couple of hundred years,' he said. 'No doubt exists that carbon dioxide contents in the atmosphere are increasing. There is no doubt that all the processes that cause this increase are escalating rapidly. There is also no doubt whatsoever that the constitution of the atmosphere has a vast influence on the amount of sunlight trapped

in the earth's atmosphere and there communicated to the earth, to
the ice caps and so on.'

It wouldn't take much, Bloom said, to push the ice caps over the
edge – 'only a few degrees centigrade'.

Bloom was a man cursed with unique foresight. He would
later carry out the first tests showing the Derwent River was
contaminated by heavy metals, prompting an intense backlash from
industries that used the river like an open sewer – just as Senator
Little had dismissed his concerns about air pollution. But some of
the committee members, at least, seemed to have understood.

When Branson went on to table his committee's final report in
September 1969, it would deliver one of the clearest, most direct
warnings about the risk posed by the greenhouse effect in Australia at
that time. In a section titled 'Nature of Air Pollution', the committee
reported that 'air pollution is basically caused by man's insatiable
demand for energy' and that 'since man first began lighting fires for
warmth and cooking, he has been contaminating to a height of from
30,000 to 50,000 feet the relatively thin layer of air which surrounds
our globe'.

It continued:

The burning of coal, oil and natural gas over the years has increased
the global concentration of carbon dioxide in the atmosphere, and
although this increase has had no recorded effect on living organisms,
it has been suggested by leading scientists that further increases
could modify the heat balance of the atmosphere. This could result
in marked changes in world climate and ultimately a warming of
the earth to such an extent that the arctic ice caps would melt and
sea levels would rise by some 400 feet. Many major cities would
become inundated as a consequence. A further possibility was that
an overabundance of carbon dioxide in the atmosphere could upset
the carbon/oxygen balance and interfere with the normal process of
plant photosynthesis, which could result in our world exhausting its
supply of oxygen. The process of contamination increased sharply
as the energy requirements of mankind increased with the advent of
the Industrial Revolution in the 18th century and again following

the general industrial expansion since World War II. Air pollution then can be described as one of the penalties of urban development and technological progress. Man has been using the atmosphere as a huge rubbish dump into which is being poured millions of tons of waste products each year. In the United States of America alone it is estimated that 142 million tons of air pollutants were admitted into the air in 1966, that is almost 400,000 tons each day.

Evidence received by the Committee indicates that an air pollution problem already exists in some parts of Australia and while not yet a problem of the magnitude existing in well known centres of pollution such as London, New York, Los Angeles and Tokyo, a problem which nevertheless warrants urgent planning and action.

Though phrased in terms of probability and not certainty, the frank declaration that 'man has been using the atmosphere as a huge rubbish dump' was stunning. Both Branson and the committee's deputy chair, Condor Laucke, were Liberals; four decades later, foot soldiers within their own party would harangue climate scientists for daring to say as much, blaming sunspots and volcanoes for fluctuations in global average temperature series and doing everything they could to kill off efforts to address the issue through the United Nations.

The report did not investigate the implications of the greenhouse effect any further, but, as a public document, it was available to anyone who wanted a copy. Though its impact would be overshadowed by interest in the Water Pollution committee's later work, its comments on carbon dioxide were debated in parliament, two of the first books on air pollution would refer to it, and the report's findings caught the attention of Australian heavy industry.

In 1971, the Australian Minerals Industry Council (AMIC), a forerunner to the Minerals Council of Australia, organised a seminar on environmental issues. Among the invitees was Kenneth Nelson, a scientist from the American Smelting and Refining Company. Unlike Bloom, Nelson was a scientist who always made sure to tell his bosses what they wanted to hear. As vice president for environmental affairs, he had once sought to surreptitiously

undermine scientific research linking cancer among smelter
workers to exposure to arsenic and sulphur dioxide – a scandal that
would only break in 2007. Because Nelson knew those responsible
for reviewing the research, he employed a strategy pioneered by
schoolyard bullies. Using trusted colleagues to launder his views,
he had them badmouth the research to the reviewers, who in turn
passed these criticisms on to the authors. The study's authors, none
the wiser, dutifully watered down their conclusions.

On the ground in Australia, Nelson delivered a presentation to
the broader resources sector on clean air, in which he went so far as
to put solid figures to the problem.

'Carbon dioxide is really the *only* contaminant which we can
say is proved to be accumulating in the global atmosphere,' he
said. 'Apparently it is being generated somewhat faster than it can
be removed. The atmospheric CO_2 concentration is currently about
330 parts per million – up from a bit under 300 parts per million fifty
years ago. No cause for alarm, but worth watching on a global scale.'

Not only had Nelson undermined any sense of urgency his
audience of anxious resource executives might have been feeling, he
had also framed the situation in a way that implied significant doubt.

Humanity was responsible for only 2 per cent of the carbon
dioxide on the earth, he went on. Despite 'frequent press reports' that
an increasing CO_2 concentration might create a greenhouse effect,
he noted that other theories suggested the impact might be offset.
One hypothesis was that particulate in the earth's atmosphere might
increase the planet's reflectivity – its 'albedo' – so that less sunlight
made it through to keep the earth warm – though Nelson freely
admitted there was no evidence to support this idea. He concluded
this part of his presentation by declaring, 'Natural processes appear
to maintain a mysterious balance.'

Humanity had observed the microbe, split the atom and landed
on the moon. The suggestion that nature was an enigma, impossible
to fully know, was interesting, coming from an alleged man of
science. If it was any indication of the seriousness with which Nelson
took his work, or the culture he operated within, it was perhaps
telling that he closed with a sex joke:

'In any case, I think we need to worry far more about our libido than our albedo.'

As the 1960s drew to an end, the Australian petroleum producers were on a collision course with destiny. The country was set to export 65 million barrels of crude oil in 1970 – a volume roughly equivalent to 215,295 suburban swimming pools, and more than four times the 15 million barrels it had exported the year before. The scale of this expansion made conflict with the early environmental movement inevitable.

Flush with cash, the oilmen closed out the 1960s on a high – and by getting better organised. The Petroleum Engineers Society of Australia was formed to help individuals working within industry form closer ties with each other as the association increasingly began to build stronger connections with its international counterparts. At one point, APEA considered hosting a joint conference in Australia in partnership with the UK Institute of Petroleum and the American Association of Petroleum Geologists, but this was soon abandoned. When they ran the numbers, they found it would be cheaper and easier just to invite a larger number of international guests to its own conferences.

It was the Great Barrier Reef caper that brought the first challenge to their legitimacy.

Lying off the coast of Queensland, the reef is the largest structure on earth built by living organisms. It took billions of coral polyps over half-a-billion years to construct the 2900 individual reefs that extend over 2300 kilometres and are home to more than 9000 known species. In the late 1960s, the whole area was still a sparkling paradise, where the infinite blue Australian sky collided with the crystal waters of an island chain. Beneath the waves was a submerged world, radiating out in smatterings of vivid blues, oranges, reds and purples. The oil explorers looked upon this spectacular wonder of nature, stuck their thumbs in the loops of their trousers, whistled at the splendour and agreed among themselves that this was the perfect place to drill for hydrocarbons.

Back on land, the quest for oil had been frustrating and financially costly, but the open seas presented a new, unregulated world of possibility. Early exploration work had identified vast potential reservoirs of oil and gas lying off the coast. The first offshore wells were drilled into the Bass Strait in 1964 by the drillship *Glomar III*; in 1967, the vessel found the Halibut and Kingfish oilfields in the area. In the 1970s, development of the North West Shelf gas field off the Western Australian coast, an ocean of hydrocarbon beneath an ocean, would guarantee the industry's survival – but that was still in the future.

All the while, the early Australian environmental movement was nervously watching events overseas. The first oil-related industrial accident came in 1967, when the tanker SS *Torrey Canyon* ran aground on rocks off the UK coast, bleeding somewhere between 94 and 164 million litres of crude oil into the sea. The stricken vessel lay partially submerged for days, dumping oil into the ocean as it broke apart, creating a slick that extended all the way to the coast of France. It was the first time anyone had encountered a disaster on this scale, and BP, who had chartered the vessel, turned to the British government for help. To solve the problem, the British Royal Navy and Royal Air Force resolved to bomb the wreck – even using napalm to ignite the spill. This, however, only made the situation worse. Tonnes of detergent had been dumped onto the tanker's deck in a futile effort to break up the spill; when it sank, the detergent was released into the water and washed ashore. It would later turn out this substance, manufactured by BP, was toxic to marine life, and the beaches slathered in it were sterilised.

There were other incidents, too. Between 1950 and 1968, there were 30 'major rig disasters' globally – two-thirds of them ending with the destruction of the rig. 'Blowouts' – the term used to describe the rupturing of an oil well, sending oil gushing to the surface in a geyser – were becoming particularly common. By the late 1960s, the industry was averaging 40 blowouts a year.

Australia would have its fair share of incidents. In December 1968, the Marlin drilling platform in Bass Strait, a joint venture between Exxon subsidiary Esso and Broken Hill Proprietary Company

(BHP), made headlines when it suffered a blowout. A crisis was averted when a Texan oil well firefighter – Paul 'Red' Adair – was flown in to fix the problem at a rate of $10,000 a day. Adjusted for inflation, this is the equivalent of more than $146,912 a day in 2024. The Marlin incident would be followed by another in August 1969, when an explosion damaged the Petrel-1 exploration well in the Joseph Bonaparte Gulf, off the coast of northern Australia.

But it was the Santa Barbara oil spill in late January 1969 that really did it for the Australian public. A blowout on a platform owned by Union Oil spilled 100,000 barrels of raw crude into the ocean over ten days, smearing beaches in South California with sludge. Images of the spill – of gulls choking on oil, their feathers thick with crude, unable to fly – spread around the world. Thousands of seabirds, fish and invertebrates along with dolphins, elephant seals and sea lions are thought to have been killed in the event. A week after the spill, witnesses were still spotting whales swimming in the oil-thickened water.

Putting two and two together, Australians started asking urgent questions about all the oil exploration being planned in the area around the reef. Queensland premier Joh Bjelke-Petersen, seeing dollar signs, had thrown open the state's entire coast to offshore oil exploration in 1968 and told the industry to let rip. Six exploration wells would be dug in the reef area as the industry drilled into paradise, but the flashpoint would be a proposal in 1969 by Ampol and Japanese state-owned oil company Japex.

Ampol was no stranger to challenges from the nascent environmental movement. Five years earlier, when it began exploring Barrow Island in a joint venture with Standard Oil under the name Western Australian Petroleum (WAPET), it had faced criticism over its industrial activity in the Class A nature reserve. To quell dissent, WAPET hired naturalist and television personality Harry Butler to conduct a survey of the area. Butler's initial work doubled the list of known species on the island, prompting international concern about the potential damage.

In response, the Explorers Club of New York – a members-only organisation for adventurous scientists – hired Butler to go back and investigate the environmental impact that exploratory drilling might have on the local environment. When Butler delivered his report, he found there would be 'no serious or lasting damage'. With those words printed in every newspaper, WAPET brought Butler on full time as a consultant in 1970 to oversee its operations. Co-opting the clout of a major conservation figure was a coup for Ampol – and is still considered by the petroleum sector to be the finest example of how to collaborate with environmentalists. What began as a few exploration wells eventually evolved into a major project today known as Gorgon. By 2001, over 200 wells had been drilled in the Barrow Island sub-basin. As of 2023, after a series of mergers, Gorgon is operated by US oil giant Chevron. It produces 15.6 million metric tonnes of gas a year.

The Great Barrier Reef caper would not go so smoothly. When Ampol and Japex announced they were bringing in a large drilling rig, *Navigator*, from Japan to Repulse Bay, just south of the Whitsundays, the Australian Conservation Foundation mobilised. Drumming up outrage, the group coordinated with the Queensland Council of Trade Unions in February 1970 to organise a boycott. The union blockade meant that once the *Navigator* hit Australian waters, it would be stranded the moment it ran out of food or fuel. Under public pressure, the Queensland and federal governments wrote to the companies and asked them to postpone drilling while they held an inquiry into the risks posed by industrial activity in or around the reef area. In March that year, the joint parliamentary inquiry between the Queensland and federal governments was upgraded to a royal commission.

APEA leapt into action. It was built for exactly this sort of moment. The organisation insisted no company wanted to drill the reef itself. Its Air, Land and Water Conservation committee rushed out a report advising oil explorers on how they might minimise environmental harms while drilling, to demonstrate to the public that they took conservation seriously, and a legal team from Minter, Simpson & Co – a forerunner to law firm MinterEllison – was brought in to represent the entire industry at the pending royal commission. To fund its legal fees, $65,000 in donations – worth roughly $890,000 in 2024 – were solicited from member companies, with nearly a third coming from APEA itself.

APEA also set up a special committee headed by Robert 'Bob' Foster to monitor the inquiry. Foster was a petroleum scientist who had been trained by Shell but would go on to serve as head of government relations for BHP. Later in life, he would become a prominent Australian climate change denier and founding member of the Lavoisier Group, an influential crop of wealthy businessmen and former politicians opposed to action to address global warming. In the meantime, he would devote considerable effort to convincing Australian governments to let guys like him drill the Great Barrier Reef.

This fight was the baptism of the Australian Conservation Foundation, and a fledgling environmental movement that terrified the oil companies. Officially incorporated in 1966, the organisation's membership included more than 350 conservation organisations. Not only did this new body threaten the oil industry's bottom line by demanding that government lock up areas like the Great Barrier Reef, protecting them from exploration, it also challenged the industry's public standing and influence by generating bad press.

Internally, APEA's thinking was conflicted: it was vital the industry address the criticism, but a public inquiry wasn't the end of the world. In fact, APEA's leadership considered the royal commission a convenient opportunity. As experts gave evidence on everything from drilling techniques to tidal flows, the technical information they shared became freely available to the public. For APEA's members, the hearings were source of new scientific and engineering data they could use on other projects, without paying a cent for this expertise.

And if any of APEA's members were heard muttering about environmental harms or seemed susceptible to being swayed, the hulk-like figure of Esso Australia's James Donald Langston would be on hand to hose down concerns.

★

Langston, an American, was the local head of the Exxon subsidiary and a bull shark in a goldfish tank. Exxon was among the biggest oil companies in the world, and its personnel were considered serious people who worked on complex projects – projects that involved big engineering, demanded high precision and generated vast financial returns. In Australia, this meant that when Langston spoke, everyone listened.

With his signature on the 1967 exploration permit for the Marlin and Barracuda oilfields off the Victorian coast, Langston operated in the years that followed with the same defiant streak that had long characterised Exxon's corporate leadership. Back home in the US, the company barely respected the authority of the federal government, and Langston had no intention of showing deference to Australian authorities in this posting. Called to appear before the senate select committee on Offshore Petroleum Resources, he was glib. When elected representatives asked him how much it cost to operate Esso's drilling project off the Gippsland coast, Langston claimed he did not know. More than that, he claimed his joint venture partners at BHP did not know, either. Despite this claimed ignorance, Langston stressed that Esso planned to 'make a profit, a handsome profit'.

'And we're not going to apologise for our handsome profit,' he said.

Langston had appeared in the country just in time – right as the Australian petroleum industry was running into trouble over the Great Barrier Reef – to take charge of the public relations response. Esso's headquarters in the US also dispatched Dr William H Lang of the Esso Research and Engineering Company from New Jersey to Melbourne at this time, ostensibly to provide 'technical support' to a petrochemical plant in Altona. In all likelihood, his actual role was political – to assist Langston in teaching the Australian industry how to talk about environmental issues.

The ailing Dr Lang was an expert on pollution and well equipped for the task. A former employee of the Pennsylvanian Department of Forest and Waters, he had joined Esso's research subsidiary in 1947. In that role, he carried out environmental studies and sat on internal American Petroleum Institute committees, making it likely he was the true author of the address, titled 'Petroleum and the Environment', that Langston delivered at the APEA national conference in Surfers Paradise in March 1970.

Opened by Joh Bjelke-Petersen and attended by 700 people, the conference coincided with the decision to upgrade the inquiry into drilling on the Great Barrier Reef to a full-blown royal commission. It was Langston's only known major public appearance at an event of this nature for which there are records. He made the most of his moment in the spotlight, hammering the Australian Conservation Foundation and laying down the industry line on environmental issues. He reminded his audience that high living standards were not possible without oil and told them that the industry was committed to conservation. The American Petroleum Institute had spent US $3 million on air and water conservation research in 1969 alone, he explained, with plans for more. More than 100 men had volunteered on steering committees to guide this research, he said. Conservation, Langston declared, was the 'watchword' of industry.

'We must consider our total environment and control its development to the benefit of all mankind. As far as industry is concerned, this is what conservation is all about – the proper use of resources,' he said.

It didn't appear to occur to Langston – or whoever wrote his speech – that $3 million spent in the US did nothing for Australia's natural environment, but that wasn't the point. It was the first time that the environment and pollution had been addressed by an industry leader and APEA figure so directly. Langston had delivered a masterclass in crisis comms when confronted by environmental issues. His talking points can still be heard from the mouths of Australian petroleum executives to this day.

He had, however, made one curious omission. Langston's address made no mention of global warming: the terms 'greenhouse effect' and 'carbon dioxide' did not appear once. Six months earlier, the senate select committee on Air Pollution had published its report directly outlining the risks posed by the accumulation of carbon dioxide in the atmosphere. The senate committee had drawn on research from the Esso Research Centre about reducing air pollution from vehicles – yet Esso's head of operations in Australia apparently had nothing to say about the risks. This silence is revealing: Langston would have been aware of growing concerns about the greenhouse effect, well known to his professional and personal networks, and so too would his offsider, Dr Lang, owing to the nature of his work.

But if Langston was trying to avoid pouring petrol on a public relations disaster, it wouldn't be long before someone else did it for him.

It was 1972 when the Australian oil and gas industry appears to have first publicly acknowledged the link between burning fossil fuels and the greenhouse effect. In this period those who gathered considered themselves 'Men of Science'. They were skilled engineers and scientists who considered themselves equipped with the skills to tackle the world's problems, so there was no fear in addressing the subject.

Against this backdrop, an internal tug of war was playing out within the industry at the time: 'town gas', made from coal, was being phased out and replaced by so-called 'natural gas', causing conflict between coal and gas producers. Australia ran on coal, but it burned dirty. Gas burned cleaner, but switching over suburban homes required a costly mass retrofit and education campaign. In this context, growing concerns about air pollution and the greenhouse effect benefited the gas producers, who argued that switching from coal to gas would buy the world more time to address the problem.

This appears to have annoyed those working for other fossil fuel producers, including RS Sherwin, a petroleum scientist with BP. At the 1972 APEA conference, Sherwin presented a paper titled, 'Energy: Major Sources and Consumption'. His main focus was on quantifying the scale of demand for energy between 1972 and the year 2000, but he also used the opportunity to take a swipe at colleagues in the gas industry concerned about air pollution, reminding them that burning gas still produced carbon dioxide. Sherwin specifically mentioned the greenhouse effect, noting it was first described by John Tyndall all the way back in 1861, before emphasising a growing demand for fossil fuels. The situation needed to be watched carefully, he said.

'What is now innocuous may, on a greater scale, soon become damaging,' he warned, ominously, and then called into question the gas industry's tactics: 'Some gross oversimplifications and distortions of emphasis spring to mind. LNG is NOT a pollution-less fuel, since it produces carbon dioxide when burnt.'

Sherwin did not make explicit that he was referring to growing CO_2 concentrations in the atmosphere caused by burning fossil fuels, but that is the clear implication. He similarly provided no data or referred to any resource to elaborate on his point. He simply accepted there was a relationship between burning fossil fuels and a changing atmosphere.

Then, having issued his rebuke, Sherwin moved on to other matters.

It wasn't until the following year that a member of the industry went on record directly addressing the question, when a chemist with a special interest in human health, Hanns F Hartmann, gave a paper at the 1973 APEA conference in Sydney.

Prior to World War II, Hanns Hartmann was a medical student at Vienna University in Austria, but his graduation was interrupted by the invasion of Nazi Germany during the Anschluss. He was on holiday in Italy when his aunt telegrammed advising that, as an ethnic Jew, he should not come back. He fled, landing in the UK before heading for Australia, where he arrived in March 1939 at the age of 23. Though he later trained as a chemist after xenophobic restrictions prevented him from re-enrolling in medicine on his arrival in Australia, Hartmann never lost his interest in the wellbeing of others.

The combination of his training in chemistry and lifelong interest in medicine resulted in an enduring professional focus on the effect of pollution on human health. At the time he addressed the APEA conference, Hartmann had been seconded as an officer of the Environmental Protection Council of Victoria; he was also co-author of a recent book on ecology and a member of the Clean Air Society of Australia and New Zealand (CASANZ). He had in fact organised the CASANZ conference in 1972, just the year before, with the financial support of BP, Esso, Ampol, Shell, General Motors Holden, BHP, Mobil Oil and Alcoa. The greenhouse effect had received top billing at this CASANZ meeting – even as the issue was ignored by the oil and gas industry in their own forums.

There is no way to know how many people turned out for his presentation at the APEA National Conference in 1973, but as the chief technical officer at the Australian Gas Association, Hartmann wasn't exactly a headliner – that honour was generally reserved for politicians and CEOs. Likewise, the title of his technical paper – 'The Role of Gas in Environmental Control' – wasn't particularly thrilling, and certainly didn't suggest what he was about to discuss.

With considerable brevity, Hartmann clearly outlined for his industry colleagues the consequences of burning fossil fuels, explaining that carbon dioxide's atmospheric concentration had risen since the turn of the century from about 290 to 320 parts per million, and what that meant.

'Recent work indicates that perhaps one third of the carbon dioxide deriving from fuel combustion accumulates in the atmosphere (0.7 parts per million per annum), while the remainder probably dissolves in the ocean,' he told them. 'Great concern has been expressed over the so-called greenhouse effect, which could cause far-reaching climatic changes. Carbon dioxide, while transparent to shorter-wave solar radiation, absorbs longer-wave infrared radiation reflected into space from the earth's surface and, therefore, acts like the glass roof of a greenhouse.'

With that, Hartmann had explained what the greenhouse effect was, how it worked, the role of the oceans as a carbon sink, the limits of knowledge at the time and the disconnect between what the empirical evidence appeared to show and what the theory of the greenhouse effect predicted should happen. He was, however, wrong about some specifics. Contrary to what Hartmann claimed, oceans and land-based ecosystems, when combined, absorbed roughly two-thirds of all CO_2. There were other technical errors, too, but it is unclear whether these were just poor writing, the work of an enthusiastic amateur or the product of a developing scientific field that had yet to benefit from the invention of satellites. What is key, is that Hartmann clearly acknowledged the connection between burning fossil fuels and growing CO_2 concentrations, and the potential for the resulting pollution to act like sleeping under a warm blanket in summer.

Hartmann concluded this section of his presentation with a qualification: although the consequences of this process 'over the long-term' were then unknown, he said, 'a close watch must be kept on long-term trends'.

At this point, the paper devolved into a discussion of the role car tailpipe emissions played in creating smog above Los Angeles, Sydney and Melbourne, leading into an argument in favour of gas as 'practically free from air pollution if its combustion is properly regulated'.

It was, however, a remarkable summary that demonstrates the extent to which Hartmann was across the early research into the greenhouse effect. Not only was he able to put solid numbers to the problem, but he referred in his footnotes to industry documents published in journals and presented at symposiums – including a collection of papers delivered at a symposium in Dallas, Texas, in 1968, titled *Global Effects of Environment Pollution*. Edited by a young Fred Singer – who would go on to become the godfather of climate denial after being denied a promotion while working for the US Environmental Protection Agency – this tome contained early works by big names in climate science of the era. It also included a report by Robinson and Robbins, the same Stanford Research Institute scientists who had warned the American Petroleum Institute about the risk burning fossil fuels posed to the climate.

Though it is not a clear smoking gun – Hartmann was not engaged in original research, and the papers he cited did not include the critical Robinson and Robbins report – his work demonstrates that information was being shared between Australian industry figures and their counterparts overseas. Hartmann had delivered his paper at the APEA conference a full three years before the

Australian Academy of Science would get around to publishing its own report on the issue, observing that human activities 'could have an appreciable effect on the climate within decades'.

Sherwin and Hartmann's APEA conference papers show the extent to which the Australian oil and gas industry already understood the danger posed by fossil fuel combustion at the dawn of the 1970s. Written by a petroleum scientist working for BP and the Australian Gas Association's chief technical officer, they are clear evidence not only that the industry understood the link between burning fossil fuels and the greenhouse effect, but also that it was publicly discussing responses to the problem.

Others, however, were already pioneering an early form of climate denial. At the very same conference at which Hartmann spoke, a presentation by Sir Willis Connolly, former president of the World Energy Conference (WEC), coolly dismissed carbon dioxide as a problem.

Connolly, who had been involved in setting up Victoria's power grid and belonged to a brown-coal industry fraternity known as the 'Barbarians', reported that an internal committee within WEC had investigated pollution issues. This committee, he said, had not bothered to consider the risks posed by carbon dioxide – 'the major product of fuel combustion' – because it was 'not generally recognised as an air pollutant'.

When I showed all these documents to Dr Marc Hudson, an energy transitions scholar and climate historian from the University of Sussex, he was not surprised. By about 1969, the potential risks of burning fossil fuels were understood. Out in the broader scientific community, work was already underway to pin down the specifics. The CSIRO had started to investigate the phenomenon, thanks to the foresight of a few particularly motivated scientists. Independent measurements of CO_2 had first been recorded in Rutherglen, Victoria, in 1971. APEA's counterparts in the broader mining industry had also started to talk about the greenhouse effect, and the issue was even beginning to seep into public consciousness. *Soylent Green* screened in 1973, the first film to depict a climate-shocked future.

What makes Hartmann and Sherwin's presentations so important, Hudson said, is that they are a historical anchor point, pinpointing the moment at which global warming 'pops up in the oil industry's own publications and materials'. They provide evidence of awareness among the members of Australia's oil and gas industry that there were risks associated with their commercial activities – and of how little their talking points have changed over time.

'This is not five months ago, or five years ago, but more than five decades,' Hudson said. 'One of the problems with thinking about climate policy is getting stuck in a perpetual present, too focused on the minutiae of who said what to whom at some recent roundtable or lobbying meeting.

'The next crucial step is knowing and being able to show how the oil and gas industry has been successfully influencing policy.'

MIDDLE PERIOD

5

Time Is an Essential Factor

The oilmen had every reason to be optimistic at the dawn of the
1970s. Even if the public service hated them – a poll by Canberra-
based consultancy International Public Relations of politicians and
bureaucrats found the bureaucracy thought of the oilmen as a pack of
self-serving hustlers whose word couldn't be trusted – the industry
could always count on its mates in high places. More than two decades
of Robert Menzies and his successors made for a cosy relationship
between the Liberal government and the oil executives – but
then, one fateful night in December 1972, something unexpected
happened. Labor's Gough Whitlam was elected prime minister, and
the world as the oilmen knew it was turned upside down.

Labor had been losing federal elections for as long as APEA had
existed, meaning the oil explorers were blindsided by the sudden
change in administration. They had no true friends in Labor, both
because they had failed to cultivate them and because elements within
the party were actively hostile to the industry. From the moment
Whitlam was voted into office, the oil executives whispered among
themselves about his Bolshie tendencies. Suddenly the issues of the
past no longer mattered as much. All the old arguments about the
appropriate level of government support for private industry gave
way to deep anxiety about the prospect of a wholesale government

takeover. Menzies had ensured the country's mineral wealth would be managed by private hands, but the way Comrade Whitlam talked about 'buying back the farm' threw the oil executives into a panic at the prospect that he genuinely meant what he said.

But it was Rex Connor, the new minister for minerals and energy, who would personify everything the oil industry feared and hated about the government. A tall, hulking figure, Connor was a physically imposing man, dubbed 'The Strangler' in the financial press for his rage. Political legend holds he once ripped a clock from a wall in parliament and threw it across the room.

Connor was perhaps more pro-oil than the oilmen, more pro-gas than the gas producers and certainly more pro-coal than the country's coalminers. Early on, he also tried to fund research into solar, but it was resources that took up most of his time. Wherever he looked around the map, he saw a country blessed with wealth – but one which never seemed able to command a fair price for its resources. Like the executives, he wanted Australia to grow rich and happy off the back of all that oil, gas and coal – and whatever else it was able to dig up. Unlike the executives, though, Connor wanted a greater share of the proceeds – perhaps even all the proceeds – returned to the Australian public.

From the get-go, Connor threw his weight around. When the heads of Woodside and Burmah Oil met with him at the start of his term, he invited them in, sat them down, then stood, thumped the table with his fist and told them he was now in charge, 'not that other sleazy mob'. His plan, according to these executives, was to buy the gas at the well-head, ship it across the country by pipeline and sell it to the Americans. Having explained this, he told them to get out.

Before long, Connor was refusing to meet with industry lobbyists and had cut them out of decision-making completely, preferring to take advice from the head of his department. He was a secretive and suspicious man who had spent many years watching how the nation's oil industry operated. Once in charge, he was determined to thwart their efforts to cultivate influence in the new administration. Right before his downfall, Connor would brag to a federal Labor conference in Terrigal, New South Wales, about how he had ended the era of influence by resource industry lobbyists on government.

Connor's greatest sin in the eyes of the oilmen was his quest to found a state-owned oil company. As minister, he handed the Petroleum Minerals Authority (PMA) a $50-million cheque and passed legislation allowing it to operate commercially as a private company. He then turned his attention to building a government-owned pipeline network that would crisscross the country. The idea was simple enough – plentiful gas in the northeast could be brought to the south for the benefit of all – but the project would prove Connor's undoing. He needed US $4 billion in loans to set up his project – roughly equivalent to A $68 billion in 2023 and to get it he turned to Pakistani commodities trader Tirath Khemlani.

Connor's first problem was the size of the loan. There were specific processes to be followed when looking to borrow on that scale – processes that would typically require a minister to work with Treasury. But with narrow-minded penny-pinchers in the bureaucracy and Treasury, Connor didn't like his chances and instead sought authority from the Loans Council to independently raise the funds.

His chosen method was another problem. Today there are financial markets and established practices for securing infrastructure fundraising from private individuals and investment firms, but at the time this was not common. A safer bet was inter-governmental lending. Connor could have gone straight to Saudi Arabia for a loan. The country had grown rich from its oil investments and was looking for places to invest, but the prospect of tying Australia to the Saudi kingdom, a major oil supplier to the US, didn't thrill him. Any such deal would have bound Australia closer to the US by proxy, compromising Connor's dream of self-reliance.

Instead, he acted on a tip. A cabinet colleague mentioned in November 1974 how he'd met a builder at a Greek ball who said he knew a jeweller who had a guy in Hong Kong with friends in the Middle East flush with oil money. The jeweller's contact in Hong Kong didn't pan out, so, after asking around, this builder put Connor onto Khemlani in London, saying he might be able to arrange a

couple of billion. Connor followed the letter of the Constitution
and obtained authorisation from the Australian government's Loan
Council to make the deal on 13 December 1974, but the scheme
unravelled as Khemlani strung him along all the way through the
rest of 1974 and into 1975. Khemlani – who some suggested had CIA
connections, and lived off potato crisps, peanuts and coffee – swore
he was good for the money the whole time. Truth was, he had never
handled this sort of deal in his life. He was in over his head and not
brave enough to admit it.

Whitlam told Connor to kill the deal and pulled his authority to
negotiate the loan, but Connor didn't follow orders. Sitting by the
telex day and night, the minister waited for news of a last-minute
arrangement that would save his grand vision. Asked by Whitlam
whether he was still trying to raise the money, Connor denied
everything. The moment presented an opening for the opposition,
who used it to whip up a corruption scandal. Coalition MP's stalked
the halls of parliament, telling anyone who would listen that Connor
had been trying to line his own pockets – but no evidence was found
to support this.

Khemlani responded by selling out. Khemlani, in safari suit and
dark glasses, flew into Canberra with eight suitcases stuffed with
documents – some written using code words – and locked himself
in a hotel room with a recently elected Liberal backbencher, John
Howard, to comb through the intel – for a fee. By November 1975,
Connor had left parliament in disgrace, Whitlam had been dismissed,
and the PMA was killed off by the courts on a technicality.

The Whitlam years imprinted themselves on the psyche of the
oilmen. It was their dark night of the soul, and the catalyst for
an ideological rebranding. No more would they play the role of
patriotic industrialists bettering their business to better the nation.
Instead, they would refashion themselves into fanatical warriors for
the free market and defenders of liberty.

But rebuilding would take time. As they looked around the
political landscape, they found little to be excited about. Fraser
was an improvement – they agreed on that – and his government
made an effort to court them, but, in their assessment, he was a soft,

squishy, small-l liberal who still didn't really 'get it'. Over in the Labor camp, a young Paul Keating was respected for his nous, but he hadn't yet totally abandoned the idea of setting up a national oil company, a policy that remained popular among his party's base. He would get there, eventually, but until then the oil industry felt they had no true friends.

As the decade wound to a close, they would suffer one final indignity: the spectacle of APEA's founding chair, the legendary Reg Sprigg, telling a room full of the nation's oil executives that solar power would one day supplant them, for the good of all humanity.

Sprigg opened his presentation to the 1977 APEA conference in Sydney with characteristic pomp. To those listening, he was a respected figure – the geologist whose work had overturned the idea that Australia was too old for oil, one of the industry's elder statesmen. He had helped start up two major domestic oil companies – Santos and Beach Energy – and was a founding chair of APEA. He had studied under famed Antarctic explorer Sir Douglas Mawson at the University of Adelaide and picked up a job with the South Australian government during World War II in which he had helped to develop the first two Australian uranium mines. ASIO had allegedly surveilled him later in his career, because of this work sourcing uranium during the war and his involvement with the Australian Association of Scientific Workers, an organisation that the nation's spooks suspected of communist sympathies. By age 30, Sprigg had discovered the world's oldest fossil in the Flinders Ranges – while on a lunchbreak – and mapped Australia's deepest undersea canyons off the South Australian coast. Even after he stepped down as APEA chair in 1966, Sprigg had stuck around to weigh in on education and environmental issues.

He had given other speeches over the years boosting the industry's environmental credentials, and his internal advocacy for conservation had finally led to the creation of a rulebook for good practice. The purpose of this presentation was to announce its release.

As a man who always spoke his mind, Sprigg didn't hold back.

He began by outlining his fundamental beliefs: that the industry had a responsibility to extract as much oil from the ground as possible, and that humanity was inseparable from its environment. These paradoxical ideas formed the basis of his personal ideology, which he termed 'practical conservation'. Next, he gave a long lecture about how the earth's climate changes across geological time, before setting out his two biggest fears: a renewed ice age, and 'peak oil', the point at which the world's most easily accessible fossil fuel resources run out.

'Many geologists believe that, omitting possible exotic consequences of man's own pollution-induced or conscious modification of climate, a return to continental glaciation could again be imminent,' he announced.

The idea of a new ice age loomed large in Sprigg's imagination. He sketched a bleak future for his audience, one in which the Northern Hemisphere had fallen under ice, agricultural regions were frozen over, nations had collapsed and war had broken out as refugees fleeing the frigid north marched south into the developing world.

Because those looking to discredit climate science would later latch onto the idea of a new ice age in bad faith, using it to cast doubt on the idea that the world was warming, it is important to put these comments in their proper context. Sprigg was dead wrong about the type of climate disaster the world would face, but his views were not entirely out of step with those of the macho world of Australian science during the mid-to-late 1970s. At the time, the emerging field of climatology was still considered a sub-branch of geology. All that maths and all those models were not attractive to hairy-chested action-men in the hard sciences who wanted to get out in the field. This culture perhaps fed a lack of understanding. In class, university students would be presented with the theory that the burning of fossil fuels exacerbated the greenhouse effect. They would next be given the data from the first global temperature series gathered by a climatologist named John Murray Mitchell in 1962, which appeared to show that average global temperatures had begun to dip around the end of World War II and continued to fall until

1960. As far as exercises went, it was innocent enough. Teaching science requires probing the limits of knowledge to show how little is actually known about a subject, particularly in emerging fields. At that point, the available evidence didn't appear to support the theory that the world was growing hotter – but that didn't discount the theory. All it meant was, anyone trying to remain faithful to the data could reasonably conclude the world might be heading for a new ice age. Those looking to eliminate this interpretation required more data.

Mitchell had updated his dataset in 1972, but two successive severe winters in North America beginning the same year further clouded the interpretation of his data, seemingly confirming the idea that the world was cooling. Sloppy reporting in the news media took the idea into the mainstream. In 1975, *Newsweek* ran a front-page story about a possible new ice age that caught the public imagination, and the idea was only reinforced when other publications followed with their own stories. It was such a powerful image that the idea would enter popular culture and circulate for decades. Later it would become central to the plot of a novel that inspired the 2004 film *Day After Tomorrow* and the Frostpunk video game.

But when climate scientists took a closer look at the temperature records, in 1976, they found something interesting. Recordings taken in North America and in some places in Europe showed average temperatures had fallen locally, but the story was incomplete. When records in the Southern Hemisphere, in places like Australia and South America, were factored in, the mercury was headed up – and only up.

Additional work clarifying the mechanisms behind previous ice ages and the specific role of particulate matter in the atmosphere would spell the end of the global cooling hypothesis – though even before that, it was a minority opinion. A review of the scientific literature by the American Meteorological Society in 2008 found just 14 per cent of articles published between 1965 and 1977 supported the idea. That review concluded that the new ice age hypothesis was 'never more than a minor aspect of the scientific climate change literature of the era, let alone the scientific consensus'. The reporter responsible for the 1975 *Newsweek* article, Peter Gwynne, even went so far as to publicly retract his work in an editorial published in 2014 by the American Institute of Physics, after climate deniers used it to spread misinformation.

In fact, a worldwide conference of climatologists held in Norwich in the UK had addressed the issue in 1975, the same year that the *Newsweek* article was published, and settled the mechanics. It was clear, they determined, that average global temperatures were going up, not down, but it took time for these conclusions to percolate through the scientific community and into public consciousness. As a result, the Australian understanding lagged. This can partly be attributed to the difficulty of sharing information in an era before the internet and partly to a failure among those teaching to keep up to date with the latest science. For the most part, those in the academy who hadn't kept up were acting in good faith, but some – like Bob Foster, the petroleum engineer who advised APEA's legal team during the Great Barrier Reef royal commission – were not so innocent. Foster would continue to misuse these early theories, taking advantage of an incomplete understanding of the data to make false claims about the contemporary science for the rest of his life.

Reg Sprigg understood, all too well, that a change in climate would have profound consequences. Unfortunately, his conclusions would be skewed by an imperfect grasp of the material. Possibly his most harmful claim was that 'man must learn to delay, modify or reverse the expected climate changes' by burning *more* oil, gas and coal.

'Fortunately, [man's] pollution of the atmosphere by carbon dioxide and other incidental by-products of his industrialisation is probably already indicating real potential for climatic modification, as by the "greenhouse" effect,' Sprigg said.

This statement was a clear acknowledgement that humanity, through the burning of fossil fuels, did have the capacity to alter a vast and complex system like the climate. Sprigg, it is worth remembering, had played midwife to the Australian oil industry; the frank words he spoke that day are evidence that Australia's oilmen understood that their business probably had a role in causing climate change. On the other hand, Sprigg was simultaneously, catastrophically wrong in his conclusions. There was no coming ice age. The reality was that pumping CO_2 into the atmosphere to forestall a new ice age was akin to collective suicide. Humanity needed to burn less fossil fuel, not more.

Not finished, Sprigg then moved on to the controversial portion of his speech: calling time on the oil industry itself. First, he laid out his belief that human progress was linked to energy use and moved through distinct stages in sequence. It was an oddly mechanical theory that suggested societies moved first from wood fires to waterwheels and then on to coal, oil, gas and nuclear power, and that we would, finally, come to rely on fusion paired with solar. No evidence was provided to support this theory, other than Sprigg's own sci-fi thinking and general observations, but his conclusions were firm.

'Ultimately, abundant solar, nuclear and fusion energy will one day take over from the fossil fuels, but time is an essential factor in the expected painful pattern of change,' he said.

He then outlined a program for adapting to climate change that would not look out of place in a Labor party platform half a century later:

'At the residential level, homes must be better insulated, leaks will have to be reduced and building designs modified more specifically to meet regional climatic demands. Solar energy research must be accelerated and the results applied widely throughout the community at both the domestic and industrial levels. Similarly, the efficiency of the motor car engine must be lifted significantly or the country's metropolitan and urban transport systems improved commensurately. These and many other innovations are already long overdue.'

If these innovations were long overdue in 1977, they would stay that way. Domestic solar hot water heaters were introduced in Western Australia in the 1950s, and some of the earliest commercial photovoltaic solar cells would be used by Telecom Australia in the late 1970s to replace diesel generators powering outback telecommunications relays. There would, however, be no significant investment in developing these technologies until at least the 1990s – and even then, it was half-hearted. Menzies may have subsidised two-thirds of the exploratory drilling process when the oil industry was getting off the ground, but the beneficiaries of this largesse would later argue fiercely, in the name of the free market, against financial support for competing technologies like solar and wind. The country, they would later say, should remain 'technologically neutral'.

Sprigg clearly left a lingering impression on the psychology and culture of the Australian oilman. The industry respected Sprigg and could not repudiate him – APEA later named an environmental award in his honour. But in delivering a doom-and-gloom vision of the future, proceeding from a flawed scientific basis to conclusions full of error, future warnings about impending risks would be waved off as nonsense. With his wild theories, Sprigg had reinforced a line of thinking among the hard-headed oilmen which caricatured their critics as 'irrational, emotive environmentalists' – otherwise well-intentioned people who didn't really 'understand' oil. When peak oil and global cooling failed to materialise in the manner described, those who followed after Reg Sprigg with warnings about global warming and climate change would be greeted with silence, or a smug, dismissive scepticism – if not outright denial and hostility.

6

One Has to Have a Programme

The convoy arrived early. It was eight in the morning and a warm winter's day in the Kimberley when police rolled up to Noonkanbah Station with a scab crew hauling drilling equipment borrowed from a CSR subsidiary. They made their final approach after a 2500-kilometre road trip from Perth, under orders from the West Australian government to break the blockade on behalf of American oil company AMAX. All along the route, people had gathered to stand on the road and block their way in support of the Yungngora campaign to stop them, but the convoy had ploughed on through.

The Yungngora had chosen Mickey's Pool Creek, a few kilometres from the station homestead, as the last line of defence. At this point the road curved around a bend and then dipped into a sandy creek crossing, acting like a natural chokepoint. In preparation for the confrontation, the community had built a makeshift barricade, working under the cover of night with their headlights off. A truck was parked across the old dirt road, two more Toyota utes behind it. A spot was left for those willing to risk being dragged off by police, with another two parked cars facing the oncoming convoy. Behind them, an additional ute straddled the road for good measure.

Two days before, the Yungngora had declared Noonkanbah

Station Aboriginal land, and the community had no intention of giving it up. The Yungngora, through the Kimberley Land Council, had previously put out a call for help, and Aboriginal people from across the region had turned out to join them. Now their efforts to resist were all coming to a head. The year was 1980. The stand the Yungngora were making here was part of a broader struggle for Indigenous land rights that had kicked off in the Northern Territory twenty years prior. Back then, the Yolgnu people from Yirrkala in north-east Arnhem Land presented federal parliament with a petition written on bark to protest the granting of a new mining lease on their land. That was followed with the Wave Hill Walk-Off, when 200 Gurindji stockmen, domestic workers and their families began a seven-year strike against Wave Hill station management and the international food giant Vestey Brothers for refusing to pay fair wages. It was the beginning of an Indigenous civil rights movement that would reshape the country.

Today it was the Yungngora's turn. A group of 60 men had waited at the blockade overnight. They included members of other local Aboriginal communities and senior Christian religious figures who joined them in solidarity. In the morning, a cloud of dust appeared on the horizon. Two men climbed a nearby hill, where they spotted the convoy, counting ten vehicles and 30 police waiting for a bulldozer to arrive.

What happened next has largely been forgotten outside the Kimberley, but at the time, this fight would command the nation's attention. The Yungngora's protest challenged not just the authority but also the legitimacy of Australia's leadership at the highest levels of government and industry. The legacy of this event, and the terror it struck into Australia's oilmen, would shape how these companies and their industry associations thought about and lobbied on Indigenous land rights through to the present day.

Like many such confrontations, it began when a century of abuse came to a head. Colonisation across the Kimberley was characterised by a particular brutality. The early European arrivals, aware they were so far from any major urban centre, behaved as if they were fighting an insurgency. Over a period of decades, Aboriginal communities were subject to punitive raids, their communities often broken up and forced off their traditional lands.

The Yungngora had been forced onto a 1700-square-kilometre stretch of Noonkanbah Station, where life was hard. The wool from the station was prized for its quality, but the human cost of production was high. Those who ran the station viewed the Yungngora as free labour. It had been that way since 1879. Early on, those who tried escaping were hunted down by police, chained to trees and burnt or flogged as punishment. As time went on, workers on the station endured slave-like conditions, as did their families. Sexual violence against women was common and the cumulative trauma was so pervasive that the term for white people, 'gadea', was used by the Yungngora as a threat to scare misbehaving children. In fact, during World War II, white people in the region suspected that the Yungngora might not resist – or might even side with – the Japanese during an invasion, because they had been so badly treated.

Conditions at the property had barely improved by the start of the 1970s. The work was hard and the days long. There were no real sanitation or laundry facilities. The community were given bare minimum rations, and the elderly and dependents were run off the station as a matter of policy. By October 1970, station management changed and the situation went from bad to worse.

It's not exactly clear which spark set off the first dispute. In 1971, an elderly Aboriginal woman died after the manager refused to call in the Flying Doctor service, but several other factors also played a role. Disputes over pay, the shooting and poisoning of camp dogs, an argument over the use of a tractor to collect firewood – all fed into a growing tension. Whatever the case, it culminated in an Indigenous exodus that was the start of a period of exile on the fringes of Fitzroy Crossing.

By 1976, Noonkanbah Station was bankrupt, and the station managers blamed everyone but themselves. According to those who ran the station, laws forcing them to pay fair wages to Aboriginal station workers were responsible, rather than the economic reality of their own mismanagement. In response, the Commonwealth government's Aboriginal Land Fund acquired the property and arranged for it to be transferred to Yungngora, who began to manage it themselves with considerable success. Two Elders, Friday Muller and Ginger Nganawilla, established a community on the banks of the Fitzroy River at the southern boundary of the station, and those who had been living in Fitzroy Crossing returned to Noonkanbah. Today it is home to 250 people who speak four different languages: Kriol, Walmajarri, Nyikina and English.

Within a week of the transfer, the community discovered that their land was subject to 497 mining leases and one oil exploration permit. It was a quick lesson in the limits of the government's generosity. Under Australian law, the Yungngora were given rights to the skin of the earth but not the mineral wealth in the soil beneath it. That belonged to the state government. It was an artificial distinction, as confusing as it was ridiculous to the Traditional Owners. As far as the Yungngora were concerned, there was no way to separate the two. Only a white man could make that mistake.

The oil company came calling about a year after the Yungngora had taken control of the station. The permit was held by US multinational AMAX Petroleum. First, they came to conduct their seismic survey; later they would seek to drill an exploratory well they called Fitzroy River-1. When seismic work began, the company received a letter informing them that their work had violated burial grounds and ceremonial areas. At first, the AMAX executives were confused: they thought they'd done their due diligence.

According to the company's version of events, the teams conducting seismic surveys had gone out to the site with Ginger Nganawilla, pointed out where they intended to work and thought they had been given the all-clear. They even had a signed document to prove it. As far as they were concerned, they had been given permission and so were taken aback when legal letters arrived accusing them of causing damage.

Events as recounted by Nganawilla to Steve Hawke, the son of future prime minister Bob Hawke and a trusted ally of the community, are somewhat different. In that version, two company men had asked Nganawilla who owned the station and then wrote his name down on some paperwork. No-one told him where they wanted to cut lines to clear paths so they could work, and no-one asked if it was a good idea. If they had, he would have told them their bulldozer was about to cut around a burial ground, tear up a visitors' meeting place, rip up a site used in male initiation rituals and plough over Friday Muller's birthplace and spirit tree. At one point, during work, the bulldozer's driver decided to take some initiative. Without asking, he attempted to improve a creek crossing but only made it worse – an incident that took place during a cultural festival and the inaugural meeting of the Kimberley Land Council.

A survey by the Western Australian Museum backed up the community's claims that the area was home to important sites, much to the annoyance of the Western Australian state government, which had given the company its full support. When the state demanded that drilling proceed, the Yungngora continued to object. Dicky Skinner, a Yungngora man sent to deliver a petition from the community to Western Australia's parliament, laid it out simply for the reporters he spoke to. 'To them, the land is just money,' Skinner said. 'All they are looking for is money. To my people the land is part of us. We don't want drilling.'

When the community, with the support of civil rights leaders like Jimmy Bieundurry, locked the gate in July 1979, the ensuing stand-off marked the first confrontation between the petroleum industry and an Indigenous community.

With time, all the predictable racist claims were levelled against the Yungngora: the community had invented the sacred sites they were defending, or were using these places to 'blackmail' the company in financial negotiations. But as the blockade went on, another, more frightening idea took root in the minds of powerful men. There was no possible way the hapless Yungngora could have decided on their own to defy the oil company, they reckoned. They had to have had help – and there was only one place they could have got it: Marxists. There was no doubt in the minds of some within government and industry that communists from 'over east' had made their way to Noonkanbah to prey on the aspirations of the community to nefarious ends. According to the conspiracy theorists, the government and the oil company weren't dealing with a land rights issue – as far as they were concerned, there was no 'land rights issue' in Western Australia to deal with. This was all the work of devious agitators and insurgents.

As Western Australia's political and business leadership were working themselves into a tizz over their own personal 'red scare', the righteousness of the Yungngora's struggle acted as a beacon to 'bleeding hearts' moved by their defiance. Some were just drifting through the area and stayed to lend support; others came explicitly to join them. Among them was Steve Hawke, and as the confrontation deepened, the Yungngora asked him to act as a first point of contact for external media.

Their demand – the right to refuse the oil company access to their land – was given more grunt when the union movement joined them. At that time, union leaders across the country saw no good reason why they should limit themselves to industrial issues, and those operating across the Pilbara and Kimberley in far-northern Western Australia were among the fiercest in the country. Wildcat strikes were common, as the local membership would often defy orders from the more conciliatory leadership down in Perth.

Confronted with an obvious injustice on their doorstep, they helped found the Kimberley Land Council, which the state government refused to recognise. When the Yungngora leadership, through the land council, put out the call for their compatriots to join them in resistance, the unions had answered.

With a moral panic in full swing, an active union movement in the mix and a militant land rights movement gaining ground, the Western Australian premier, Charles Court, treated the blockade as a direct challenge to his authority. In the premier's view, the situation couldn't have been simpler. As far as he was concerned, 'the land of Western Australia does not belong to Aborigines [sic]'. Oil companies applied for permits, and the government decided whether it would grant them. Once granted, that was it. Court considered any intervention in that process as compromising the authority of the state government.

Court demanded that the company go ahead and drill, and the unions responded by voting on 25 March 1980 to call a boycott in support of the blockade. The decision had the blessing of the ACTU, then headed by Bob Hawke, meaning that the drilling operation would have no logistical support. Without trucks, fuel or drivers, the AMAX drill rig and other equipment sat stranded in Perth.

Over in Canberra, headlines about the escalating situation drew the attention of Prime Minister Malcolm Fraser. Rather than deal directly with the Western Australia premier, Fraser went directly to the AMAX board in the US to talk them down. When they agreed to withdraw from the project, Charles Court was furious. When he learned of Fraser's involvement, he was enraged.

The premier simply repeated his demand that the company start drilling, but this time he threatened that if they failed to carry through, he would stop them from ever getting another chance to explore for oil in Western Australia. Court later explained his reasoning to his biographer, Ronda Jamieson, saying: 'If we walk out of Noonkanbah it will be the next station and the next station, and the next, and the whole of Western Australia will be denied oil exploration.'

Next, Court responded to the union blockade by setting up a government-owned trucking company called Omen Pty Ltd, formed with help from a law firm at which the president of the WA Liberal Party was a partner. He staffed the firm with scab drivers, and a convoy of 49 trucks, bound for Noonkanbah, left a Midland Brick yard – also owned by Liberal Party donors – in the dead of night, with a full police escort. The drivers wore fake beards, for fear of union retaliation.

At multiple stages, the Yungngora and their supporters made creative attempts to block the route in protest. One group of 100 people stopped the convoy at Tabba Tabba Bridge, west of Strelley, when they covered the road with rocks and stones. Further up the road, a picket line of 400 people was set up in Broome, but the convoy ploughed through at 30 kilometres an hour with sirens and horns blaring.

The final challenge came that morning at Mickey's Pool Creek. Slowly, the police moved each vehicle blocking their way with a bulldozer, until they were confronted with a group of men who stood defiant. To begin the assault, the police ordered the Aboriginal police aides to the front ranks, some of whom were being asked to arrest relatives. As people were dragged away, Ginger Nganawilla began a slow, chant-like song. One by one, those present joined in, until only the white religious ministers, who didn't know the words, remained silent.

Having arrested everyone, cleared the blockade and arrived at the site, the convoy would face one final challenge, this time from the crew of 20 roughnecks who were supposed to run the rig. They were members of the Australian Workers' Union, and it was not at first clear if they would join the blockade or break it. It took some time, but they too refused to work the drill-site after the ACTU, acting on a direct promise made by Bob Hawke, gathered $128,000 in donations to cover their wages. Their refusal forced the government to organise another scab crew to run the rigs.

When AMAX began to drill on 29 August 1980, the entire Noonkanbah community turned out to bear witness. Under the Yungngora's gaze, the oil company's scab crew, with their police escort looking on, punctured the earth. In the end, it would all turn out to be for nothing. The well proved dry; there was no oil at Noonkanbah.

As all this was going on, APEA had been busy. Marxism was in the headlines in May 1980, AMAX was in trouble, and the prospect that a group of Indigenous people might get a real say over what happened on their land struck fear into Australia's oil executives. Confronted with these problems, the execs decided what they needed was a little intelligence.

The man they turned to was Leonard 'Len' Stanley Barker, an ex-journalist turned spinner with his own firm, BPR Pty Ltd. Barker had been involved in oil exploration around Mereenie, in the Northern Territory, which is likely how he came to be known by the oilmen. It was Barker who initially approached APEA with a proposal to monitor and report on the budding Indigenous land rights movement. In his letters, Barker sought to assure APEA they had the right man for the job. He had 'extensive experience' dealing with Indigenous communities, he explained: he had once worked as a pearl producer in the Torres Strait and Papua New Guinea, 'during a period in which [Indigenous communities] were aspiring to "better things" without knowing what those better things were (outside of more cash in hand).'

In going about his task, Barker would be given a free hand. What APEA wanted was a state-by-state analysis of land rights legislation, an explanation of the basics of sacred sites and a close eye kept on the activists working in support of the Yungngora. In their own words, they also sought to know whether sacred sites were 'being cynically manipulated to create obstacles to exploration or to reinforce additional financial demands', 'the extent to which the Aboriginal Land Rights movement is manipulated by national and state groups' and 'the extent of foreign political involvement in fostering agitation'. Barker was given strict instructions to remain in regular contact with APEA's representative, Ted Garland of Alliance Oil Development Australia.

Barker kept extensive records of his work over the course of his 18-month association with APEA; one folder in which he kept records of his research was labelled 'Abo's running notes' – a casual use of a racial slur indicative of attitudes widespread among the Australian community at the time. In his letters back to the organisation, he reported that he had engaged 'a total of eight people' to fill various positions, including a researcher he claimed to have embedded with the activists themselves. The industry association didn't appear to have asked for this special service, but there is no record of any objection from them when Barker informed them that his informant had 'spent 15 days in the Kimberleys during the height of the Noonkanbah crisis, living and travelling with some of the key activists'.

That researcher was Joseph Poprzeczny. As Poprzeczny tells it, he had been introduced to Barker by 'a friend who worked in Canberra' after he had taken an interest in events at Noonkanbah. Barker then asked him to write up his thoughts. Later, realising something significant was happening in the area, Poprzeczny says he asked to travel to the Kimberley as a researcher to 'fine-tune' his thinking.

Poprzeczny appeared to have a fascination with progressive activist movements; in one of his reports to Barker, he mentions that he had once 'passed across a lot of information about the role of student unions in relation to anti-development movements' to geologist John Grover, a right-wing propagandist who published two 'exposés' about early environmental movements. In 1977, Grover had been paid by resource companies to run a pro-nuclear letter-writing campaign, calling on the government to open up the country to uranium mining. It was not successful but, with the benefit of hindsight, Grover was a pioneer: he appears to have set up an early pen-and-paper version of a social media troll farm.

In his letters, Poprzeczny was clear about his own views, telling Barker that the 'sacred sites movement' was a 'monstrous phenomenon' in a democratic society and describing it as an extension of the anti-nuclear movement, which he saw as 'anti-development'. As far as he was concerned, he says, he just wanted to see what was going on.

How Barker thought about his researcher's activities during his travels is another matter. Not only did he use Poprzeczny's trip to advertise the lengths he would go to in service of his clients, but Barker also gave his researcher strict instructions to keep his involvement 'highly confidential', as he 'did not want to see this in a book'.

In early August, as the convoy was on the final leg of its run to Noonkanbah station, Barker gave Poprzeczny his brief. He wanted his researcher to look into reports of racial violence in the area and the potential for 'black militancy'; to assess local politics and the role of local religious figures in influencing the course of events; to gather background information on key activists; and to report if there was any sense the Yungngora might be growing tired of their non-Indigenous advisers.

It is not clear whether Poprzeczny actually made it onto Noonkanbah before the dispute reached its crescendo, though he reported having watched police escort the drill rig to the station. Years later, when I showed him copies of his reports and correspondence, his answer was vague.

'I made a longish visit to the Kimberley and visited several places, I recall, even a station out on or near Bonaparte Gulf, or up that way,' Poprzeczny said. 'So, a bit of fishing and sightseeing, including of herds of wild donkeys. A very long and interesting trip, during which a mate up there was generous with his time and vehicle.'

Over the course of the project, Barker would produce three reports for APEA. In a timeless tradition, Barker appears to have taken credit for Poprzeczny's work. Correspondence between the two appears to show that Poprzeczny was the true author of the first two reports but this contribution went unacknowledged.

The initial report, which provided a foundation for the others, was based on what its author called 'an elementary understanding of aboriginal religious beliefs/mythology/law', which had allowed him to 'form an opinion whether "Noonkanbah" is likely to be repeated'.

The people depicted in this report seem to me racist caricatures, consistently attributed the worst possible motivation and the events described are viewed through a conservative Christian lens. The report began by speculating about the 'apparent history of the aboriginals living at Noonkanbah'. It suggested the Yungngora belonged to 'the Walmadjeri tribe who infiltrated or migrated to the area from the Western Desert over a period of seventy or more years'. This migration was attributed to a clash within the community over 'mytho-religious complexes' partly involving a 'cult connected with evil spirits and black magic'.

The Walmadjeri, it claimed, had gone on to 'dominate numerically' and culturally 'conquer' the Kimberley region, with 'violence on a small scale' erupting where other Indigenous communities sought to resist a process that was described as 'Walmadjerization'. In seeking to explain how a sacred site 'may emerge today in a place it did not exist yesterday', it acknowledged these places could emerge from the ongoing practices of Indigenous culture, but added it was also possible their existence and location could be 'conveniently manipulated' by 'some aboriginals and their white advisors'. The term 'sacred sites', the report warned, was 'dangerously misleading' and had been 'stretched beyond recognition in the course of the Noonkanbah struggle'.

Dismissing other, more objective and sympathetic descriptions of Indigenous culture as 'idealistic', the report focused on 'key aspects of their religion such as that sex, reproduction and excrement figure largely in their mythology and rituals, that superstition with its essential companions, fear and black magic, also figure prominently while swift and remorseless punishment, executed by the dreaded "Kadaitcha" men, followed any transgression of the law'. Unlike 'Christian, Moslem or Hindu religions', the report said, Indigenous religion was not 'static'. A more 'cynical' view, the report argued, would be that Indigenous spirituality was actually 'opportunistic'. 'Primitive' Indigenous belief had, it claimed, incorporated 'elements of Christianity', leading to the creation of what the report claimed was a 'cult'.

More than that, the report lingered over lurid descriptions of male initiation ceremonies in specific regions of the Kimberley, commenting on 'the fervour which drives a male to submit to the ritual of circumcision and particularly sub-incision without the benefit of anaesthetics or antibiotics'. For good measure, it noted how 'some white advisors and associates of aboriginals have undergone sub-incision in order to gain initiation into a tribe'. This line was printed in italics.

Whether or not any of this had basis in fact, the scene was set. Having presented Indigenous people of the region as naive and potentially dangerous savages who struggled to make decisions on their own, the report laid out a timeline of events at Noonkanbah. The initial report acknowledged there were genuine concerns among Indigenous communities about their historical treatment, but it broadly sketched the Yungngora as misled by progressive Christians and radical lefties who had parachuted in from over east. The report poured cold water on the idea of communist infiltration, but also noted the involvement of one activist whom it speculated might attract a financial donation from Libya's Muammar Gaddafi. In a postscript, the report also informed its readers that a Western Australian academic had reported that 'some three years ago' a

Chinese researcher had sought 'an itinerary for meeting Aborigines in the state'. The apparent implication was that the Chinese Communist Party marched behind him.

To provide more detail about the activists, Barker forwarded Poprzeczny documents from the office of 'B.A.S'. This was almost certainly BA Santamaria, an ultra-conservative Catholic reactionary who had been keeping records on some of these individuals. Poprzeczny, however, rejected them as 'unnecessary' though 'interesting'.

The follow-up reports built upon this work but were mostly concerned with the developing legal framework around Indigenous land rights. Poprzeczny took the liberty of giving Barker, and by extension APEA, some free political advice in a letter that accompanied delivery of the last report. In his view, the oilmen should 'encourage Liberal Governments' to set up a system of individual private property for Indigenous people across the country.

'The Oil Industry, like the Jimmy Bieundurry's and others have to have a clear cut goal. And to have that goal one has to have a programme,' he wrote. 'And if that programme falls short of individual private land ownership, then you have problems; room for Churchy activists.'

Through it all, Barker and Poprzeczny showed an odd fixation on Steve Hawke. In one letter, a handwritten note had been added describing Hawke as a 'very nasty little boy' – though it is unclear whether it was written by Barker or Poprzeczny. When I showed Hawke sections of the reports and letters in which Poprzeczny discussed his activities and those of his friends, he was amused. Poprzeczny, Hawke said, had 'some things right, some things severely mixed up and a whole lot of things just plain wrong'.

'What a very strange experience it was reading back through the ravings of mad conspiracy theorists of forty years ago,' he said. 'What they cannot give credit to is the fact the blackfellers had decided to stand up for themselves after a century of being ground under a variety of heels.

'The absolute apparent injustice and righteousness of their cause attracted all sorts of bleeding hearts, from naïve teenagers like myself to old Christian lefties like Stan Davey.

'We all just happened to be there at the time through a variety of happenstances.'

For all this fine work, APEA paid $27,780 over two years, the equivalent of $141,720 in 2024 dollars.

The day before AMAX broke ground in Noonkanbah, Fraser summoned representatives from APEA and the Australian Mining Industry Council – a forerunner to the Minerals Council of Australia – to yet another meeting. There had been several in this period, but with the blockade broken and drilling actually about to begin, this one was about setting up 'ground rules' to stop another Noonkanbah from happening in the future.

Barker may not have been the sharpest operator, but for better or worse APEA would rely heavily on his intelligence in these encounters. Meeting minutes and internal briefing documents in Barker's files reveal that the oil companies were nervous. Esso and Shell both had plans for other developments in northern Australia, and concerns were growing that what they perceived as Indigenous militancy might spread through the region like a contagion. The broader land rights struggle was already being viewed as a threat because of the potential for Indigenous communities to gain the right to control what happened on their land, including the right to veto resource exploration and extraction. Barker's files appear to acknowledge that the reverse was also true: if the legal frameworks introduced by Whitlam covering land rights were repealed, the total uncertainty it would create meant 'all mining would stop'.

Ideally, what the oilmen wanted was to be left alone to deal with Indigenous communities one on one, without 'interlopers'. This had proven 'effective' in the past; the asymmetry of resources meant they had the advantage in any negotiation. In one meeting, Esso's Ken Richards explained that the company 'would not have been able to proceed' with one particular project if the local Indigenous people 'thought their operation was a mining one, such as an open cut'. At law, state governments owned the resources beneath the earth; once a licence was issued, permission to drill had been granted. That was it. The oil companies held all the cards, and so long as they were polite about it, they found they could do what they would.

The minutes also record that the industry 'fundamentally' agreed with the position taken by Charles Court at Noonkanbah, even if the brutality of his stand had created a public relations nightmare. The oilmen agreed among themselves it was a 'pipe dream' that state governments would accept a uniform federal approach to land rights – but there were also worries that if they were wrong, they might be made to pay compensation to Indigenous communities for damage or disturbance caused during exploration. Going into these meetings, everyone agreed not to bring up the matter with the minister, apparently in the hope it simply wouldn't come up.

No Indigenous organisations would be present at the meeting when Deputy Prime Minister Doug Anthony told the APEA representatives he wanted to 'establish the basis for exploratory talks to evolve a system to avoid Noonkanbah-type disputes in the future'.

'We are trying to establish ground rules under which the "mining" industry can operate,' he said. 'If Noonkanbah got out of control, the pressure on the Commonwealth to use its powers would be immense, and that is why we are trying to establish ground rules.'

The 'situation' was being 'left wide open to radicals throughout Australia', he explained, but the government wasn't going to 'sit back and let it happen'.

The task of writing these guidelines would fall to APEA's incoming executive director, Keith Orchison, and Esso's head of public relations, Frank Hooke, who had been monitoring what was happening in the Kimberley for the Exxon subsidiary. Early planning was underway on an exploration project near Point Torment, and Esso held permits in the Timor Sea. Negotiations were also on foot with the Northern Land Council in the Northern Territory to build an emergency helicopter fuel dump on an island in the Timor Sea. If not managed correctly, the feeling was that Noonkanbah could be the small domino that set the other dominoes falling.

In early 1981, Hooke flew into Perth to consult the state government about guidelines and stopped in Adelaide on his return to link up with Orchison on the sidelines of that year's APEA

conference. Together the pair met with AMIC director James
Strong. During this meeting, Strong proposed a voluntary code of
conduct to head off any push for uniform federal rights legislation.

Following the Adelaide meeting, Orchison pulled together
all the intelligence APEA had gathered, along with additional
background information gathered by oil companies with projects
in northern Australia, including Esso, Magellan and the Associated
Group. Off the back of this work, he wrote out a set of guidelines
titled *Aboriginal Communities and Petroleum Exploration and
Development: Advice for Members of the Australian Petroleum Exploration
Association Limited.*

It was a classic public relations manoeuvre: either the industry
offered up a way forward, or they risked being overtaken by events
and having a decision imposed upon them. The covering letter
outlined APEA's basic position before encouraging patience and
consultation with Indigenous communities 'as a matter of courtesy'.
Its 'principles' were more a general checklist of things companies
were advised to think about before embarking on a new project,
including the consultation process, local employment, community
development projects and the need to take care around sacred sites.

A closer read shows that the guidelines were geared to help
facilitate continued extraction. Essentially, they advised companies
on how to operate with good manners. Orchison emphasised the
complexity of Indigenous culture and sometimes sharply different
concerns of First Nations people within and between regions,
but he also stressed that 'there is no totally traditional Aboriginal
lifestyle being practised in Australia today'. He then blamed 'the
failure of many whites involved in this issue to recognise the range
of Aboriginal communities and their differing needs' for 'much of
the confrontation and confusion experienced in recent times'. At
their core, the guidelines encouraged the *performance* of consultation
even as they made clear that a company had the legal right to take
what it wanted once a permit was granted.

'As an association, APEA seeks an acceptable balance between
the interests of all involved in an area attracting exploration activity,'
Orchison wrote.

Who would judge whether an 'acceptable balance' had been struck?

The oil companies, of course.

The draft guidelines were initially shared with the Western Australian state government for feedback. The minister for resources development, Peter Jones – a man with a penchant for colourful insults – wrote back to APEA accusing them, in short, of going soft. The WA state government had staked out a position that was as simple as it was uncompromising: it would decide where explorers could look, and then, when permission was granted, it expected a company to move in and drill. End of story.

Orchison, however, was determined to thread the needle and sent a copy of his guidelines to the deputy prime minister's office for review. When a meeting was called to settle the issue, Orchison realised Anthony had not actually read them. According to Orchison's retelling of events in an authorised history of the association's politics, Aboriginal affairs minister Fred Chaney was arguing for a stronger response by the federal government, but halfway through Chaney's appeal the deputy prime minister was called to another unrelated meeting with Prime Minister Fraser. Sensing his moment, Orchison also excused himself to go and wait outside the prime minister's office, where he and the deputy PM's senior adviser, Alan Hayman, 'collared Anthony as he came out of the PM's office and thrust the paper into his hands'.

As Orchison told the story: 'He admitted he hadn't had time to read it, but when Hayman handed it to him he read the executive summary there and then and quickly got the drift. He said, "I think we can go with this," and we went back into the meeting.'

When the meeting resumed, according to Orchison, Chaney realised he had been outmanoeuvred.

'A rose by any other name,' Anthony allegedly told Chaney. 'The PM and I think this APEA recommendation is a better way to go than legislation.'

It was a fun story, recounted in a book celebrating APEA's 50th anniversary, but its boasts of Orchison's skill as a lobbyist don't necessarily stack up.

There is no question APEA was lobbying on land rights beyond Noonkanbah. In its 1984 annual report, the organisation described 'onerous Aboriginal land rights legislation and environmental restrictions' as 'one of the greatest impediments to exploration', saying, 'APEA regards the settlement of the land rights controversy as fundamental to the industry's capacity to fully explore the country's onshore petroleum potential.'

But when asked about the incident, Chaney – a respected Aboriginal affairs minister, who would later describe the brutality at Noonkanbah as a low point in his life – said he could not remember any such meeting with Orchison, or Anthony's alleged coup de grâce that supposedly signalled his defeat. In addition, Orchison and the minister appeared to share a good relationship; in 1989, Orchison would invite Chaney, by then in opposition, to speak at that year's APEA conference.

More than that, the timeline didn't add up. In November 1980, Chaney had become social security minister, so he wouldn't have been in the room for a meeting with Orchison in 1981. His successor as Aboriginal affairs minister, Peter Baume, also had no memory of meetings with APEA during his tenure and claimed to have never once interacted with the organisation while in the role.

Orchison, it seemed, had been bullshitting when he spoke to history.

7

No Regrets

The Australian oil industry had never truly been concerned about the opinions of the general public, but at the start of the 1980s they discovered just how much they were hated. As they looked to rebuild after the tumultuous Whitlam years and the frank admission by APEA's founding chair that the industry was on borrowed time, the association commissioned Geoffrey Luck of Roy Morgan Research Centre to find out what actual Australians really felt. When the results came back, the advice – in so many words – was that people thought the oil companies were a pack of profiteering cutthroats dominated by American financial interests.

Luck suggested that if the oilmen wanted to mend their image, they shouldn't bother talking to the population at large. People couldn't tell the difference between an oil explorer, a refiner and a retailer, and weren't particularly interested in learning it before writing them all off as bastards. He instead suggested they were better off directing their efforts towards making friends in politics – specifically the government of the day and the Department of National Development. To do this, Luck advised the industry to cultivate relationships within the bureaucracy, stick to the facts and focus on making targeted policy submissions. For good measure, he pointed to the internal division that had grown between the big multinationals and the small explorers within APEA and politely suggested they pull themselves together.

Over the next two decades, that's what the industry set about doing. As head of the organisation, Keith Orchison would be remembered for having the ability to think ahead by several decades, marking him out among other industry figureheads. Orchison, a South African and journalist by profession, tended to take the long view. Some would go so far as to call him 'visionary'.

He was also an unusual hire in that he wasn't an oilman. He had been working in the public relations department at La Trobe University and editing a trade publication when he interviewed for the position of chief executive. During the conversation, he allegedly made no secret of how little he thought of the organisation. APEA hired him anyway.

What he brought was a sense of pizazz to a stuffy industry body concerned mostly with the technical and financial aspects of its business. A sharp communicator, Orchison repeatedly demonstrated his worth in his first few years with the association. In the industry history written for the association's 50th anniversary, Orchison told how in 1983 he booked Prime Minister Malcolm Fraser to open the conference in Melbourne that year, but the arrangement fell through when Bob Hawke was elected prime minister days before the event. In search of a replacement, Orchison allegedly phoned Mal Bryce, deputy premier of Western Australia, at 11 pm on the night of the election, a Saturday, and asked whether he could fly to Melbourne to open the show first thing Monday. When Bryce hesitated, saying that he didn't have a speech, Orchison told him not to worry – he would write him one. When Bryce spoke, he did so with the industry's words in his mouth.

Orchison took the job at a difficult time for the global petroleum industry. The price of oil was falling and demand was dropping domestically. Producers were pottering along, but Orchison's counterparts at the Australian Institute of Petroleum (AIP) complained in their annual report about a year of 'dismal' profits. On a positive note, AIP's CEO suggested the losses its members were taking proved wrong the 'caricature' of the oil industry as 'possibly involved in covert collusion to maintain excessive profits'.

By this point, the oil industry's understanding of the Australian political landscape appears to have solidified, categorising people based on their relationship to oil. Those most trusted by industry either worked in it, used to work in it, or had grown up around it. The next tier down was the industry's extended family – those active supporters in government, politics and media who made an effort to court the oil companies for reasons ranging from patriotism to self-interest. Below this group were the wishy-washy centrists who might not have liked the oil companies but who would occasionally fall in line, depending on the merits of a particular case.

On the lowest rung were the industry's 'opponents', and it didn't take much to qualify. The people lumped into this group were considered 'not serious' and were automatically assumed to be acting in bad faith. They included lefties, conservationists and activists – all of the 'emotional' people who, in the industry's view, did not 'understand oil' and who were potentially funded by suspicious foreign organisations, or worse, the government. After the Franklin Dam blockade, the industry even had a name for the worst among them: 'greenie'. This was the most vicious insult the oilmen could level at their enemies and the ultimate slur against someone's character. All it took to be tagged as a greenie was a fondness for nature, showing concern about whale populations or questioning the role of CFC gases in punching a hole in the ozone layer over Antarctica. Within company culture, the term was used to police the workforce: those asking too many uncomfortable questions might be warned off with the suggestion that they were 'sounding an awful lot like a greenie'. When confronted with a greenie out in the wild, the oil industry omerta applied: no engagement was permitted – greenies should be kept as far away from decision-making processes as possible, lest they gum up the works with their incessant hyperventilating about the environment.

It was a way of thinking about the world that would only harden as the industry entered yet another pivotal moment – the moment when the global oil industry realised their enemies would one day come for them.

★

In 1984, the fossil fuels humanity burned for heat, food and transport generated 205.35 billion tonnes of CO_2, and the global industry was growing worried that the public would soon wonder what this was doing to the atmosphere. Environmentalists were giving the industry so much grief that Exxon's leadership wanted to get ahead of potential trouble by coming clean to its fellow oil companies. Early moves were already being made to scrub lead from petrol and close the hole in the ozone layer through international cooperation. It was only a matter of time, Exxon's leadership reasoned, before people put two and two together and turned their attention specifically to the risks posed by burning oil, gas and coal.

So it was that Exxon gathered together representatives of the world's biggest oil companies at an IPIECA meeting in Houston, Texas for a presentation. Established in March 1974 at the request of the United Nations Environment Programme, the International Petroleum Industry Environmental Conservation Association later rebranded as IPIECA, adopting its acronym as its name. As a shadowy technocratic institution, it had limited brand recognition among the public, especially in distant Australia. But in European centres of power like Brussels, Paris and London, its word carried weight.

Initially, IPIECA's remit was to act as industry liaison, giving input to the United Nations on subjects like managing oil slicks and seismic testing, but from roughly 1984 onwards, the world's oil companies would increasingly look at the institution as a convenient organising tool. Exxon's message to the global oil industry during that presentation in Houston was clear: the 'greenhouse effect' – always rendered in scare quotes – was *actually* real; CO_2 was *actually* increasing in the atmosphere at a record rate; the climate would *actually* change; and the oil industry had better prepare to defend itself.

Bernard Tramier, environmental director of French oil company Elf, gave an account of this meeting to French researchers:

The moment I remember really being alerted to the seriousness of global warming was at an IPIECA meeting in Houston in 1984.

There were representatives from most of the big companies in the world there, and the people from Exxon got us up to speed. [...] They had remained very discreet about their own research [on global warming] [...] Then in 1984, perhaps because the stakes seemed to have become too great and a collective response from the profession required [that] they shared their concerns with the other companies.

Even this brief account doesn't quite capture the true level of Exxon's awareness. Journalists and researchers have spent years investigating what the company knew and when they knew it. The documents they uncovered have since been made public and are freely available to anyone who wants them. When Canadian journalist Geoff Dembicki sat down and read them all for his 2022 book *The Petroleum Papers: Inside the Far-Right Conspiracy to Cover Up Climate Change*, he was able to trace the development of both Exxon's early research programs and the disinformation machine the company built to hide the implications of what it had found.

Attempting even a brief summary of what Exxon knew and when is not easy. One possible starting point is the internal report written in 1970 in which an Exxon-owned subsidiary, Imperial Oil, first identified carbon dioxide as a pollutant. Another is the 1978 briefing that senior company scientist James Black gave to Exxon executives about the state of climate science, in which he directly identified the burning of fossil fuels as contributing to the greenhouse effect and warned that the risks would be immense if it continued. Black even went so far as to sketch out different scenarios for what would happen if the world continued to produce CO_2, advising that 'it is premature to limit use of fossil fuels, but they should not be encouraged'. Following this, Exxon increased its research funding into the 'potential greenhouse effect' in response to pressure from environmental groups. Its goal was not noble but rather to stay ahead of its opponents on key issues and prepare for possible litigation. The company even allocated US$1 million in 1978 – roughly the equivalent of $7.9 million in 2024 dollars, or A$12.2 million – to retrofit a supertanker as a floating research lab.

This vessel would sail from the Gulf of Mexico to the Persian Gulf, trawling for data to assess how much CO_2 the ocean could absorb.

By 1980, the company had precise figures: as of 1978, CO_2 concentrations in the atmosphere had risen to 330 parts per million, up from 280 parts per million at the start of the Industrial Revolution – almost the same figures Ken Nelson had quoted at the AMIC conference in 1971 and Hanns Hartmann had quoted in his paper for the Australian oil industry in 1973. When researchers later reviewed Exxon's projections from this era in a study published in *Science* in 2023, they found its scientists at the time were so skilled, they had forecast the effects of climate change precisely enough that their predictions aligned with contemporary models. They had even dismissed the idea of global cooling.

Another memo from around this time, written by Exxon research manager Roger Cohen, makes clear the company was aware of the risk of harm. Within the company, the scientists had been debating the implications of what they were finding as their research went on. When a colleague suggested that it might not turn out so bad, Cohen fired back, firmly hosing down the suggestion. He explained how climate impacts by 2030 might fall 'well short of catastrophic', but that didn't mean humanity had dodged a bullet. 'It may well be possible,' he warned, that continuing to burn fossil fuels 'will later produce effects which will indeed be catastrophic (at least for a substantial fraction of the earth's population).'

All of this culminated in that fateful 1984 meeting of oil company figures in Houston, setting in train a series of events that would lay the foundation for a global campaign to stymie every effort to address climate change – a war in which IPIECA would play a key role. As an official UN institution, it possessed legitimacy and authority. By virtue of its status, it plugged the oil companies directly into official processes and allowed company executives to attend world events without having to identify which firm they worked for. This would prove an advantage as world governments began to organise climate summits through the UN Framework Convention on Climate Change (UNFCCC). With Exxon taking the lead, IPIECA functioned as a convenient coordinating body, working alongside other groups like the International Association of Oil & Gas Producers.

There is also circumstantial evidence to suggest that the global oil industry kept their Australian colleagues in the loop following Exxon's presentation. By 1983, there was already a growing cultural awareness of the problem after *The Australian* published a story about climate change on its front page. The same year Exxon came clean, Australian Institute of Petroleum (AIP) executive director Peter Parkin, flew to London to attend the annual IPIECA conference. According to AIP's 1984 annual report, Parkin also 'conferred with the UK Petroleum Industry Association, the Institute of Petroleum and, in the U.S.A., with the American Petroleum Institute' – all major oil industry bodies which would have been represented at the meeting in Houston and been familiar with Exxon's research.

That year AIP was beginning to coordinate more closely, not only with its counterparts overseas but also locally with allied industry groups. Back at home in Australia, Parkin recorded how he had formalised a relationship with the Federal Chamber of Automotive Industries (FCAI), the peak body for car manufacturers, by creating the Automotive and Petroleum Industries' Consultative Council (APICC).

As AIP was hardwiring itself into international and domestic industry networks, APEA similarly announced it was partnering with the Australian Gas Association to measure CO_2 and methane emissions in their members' operations. A reporting system was developed by an internal committee known as the 'Greenhouse Technology Advisory Group' for all petroleum exploration and production operations in Australia, including action plan reports and methodologies. This allowed APEA to control how information was collected and what was presented; if asked, it could always share this data with government. What might be considered a helpful favour also afforded them obvious advantage. Whatever the data said, what mattered was that their people controlled the methods of collection, processing and presentation of this information. The arrangement would carry on until 2008, when the federal government required industry to gather this information in a standardised way for collection.

If Australia's oil industry had been made aware of the looming risk of catastrophic climate change, it didn't share this information with the public. APEA busied itself with various fights over conservation issues, and its counterparts at AIP were focused on the phasing out of leaded petrol – an encounter with government that proved good practice for AIP's later work on climate change. The same talking points found in AIP pamphlets of the early 1980s generating doubt around the science of leaded petrol would echo those later used to create doubt about 'the greenhouse effect'.

All the while, awareness was growing among the public. Research teams were already working on independent studies of the greenhouse effect; scientists had long documented their findings in academic journals; news organisations were reporting on the developing science. Even intelligence bodies were giving the issue close scrutiny. As early as 1981, the Australian Office of National Assessments (ONA) – an intelligence agency set up by former Prime Minister Malcolm Fraser to consider long-term security issues that fell outside ASIO's remit – investigated the threat posed by the greenhouse effect to the nation's fossil fuel exports. It correctly concluded that the 'potentially adverse implications' of the 'carbon dioxide problem' would 'arouse public concerns and so engage the attention of governments'.

The document first came to light when it was handed to Professor Clive Hamilton in a stack of materials during research for his book, *Scorcher*. The origin of this report was thought lost to time; for years it has been unclear if Fraser was ever presented with its contents. The answers, it turns out, lay with former CSIRO scientist Roger Gifford who says it originated with a briefing he authored for the ONA on the subject. As a plant physiologist working in the Division of Plant Industry, Gifford's research looked at how CO_2 interacted with plant life directly through the CO_2 fertilising effect and indirectly through atmospheric warming. Working with Graeme Pearman's group in CSIRO's Division of Atmospheric Research earned him a reputation as an expert in the effect of CO_2 on plants. One day, Gifford recalls, a man appeared in his lab. This man introduced himself as an analyst working for the ONA. He went on to explain that 'Malcolm Fraser had asked the ONA to prepare a report on the greenhouse effect that would inform cabinet on the subject' and that 'he had been tasked with finding out about it and writing a report'.

Gifford couldn't immediately recall the name of the analyst, but does remember him as an 'affable fellow' and 'very chatty'. He had no trace of an accent but did inform Gifford he was Welsh, adding that his friends nicknamed him 'Dai the Spy' when they found out he worked with ONA – 'Dai' being short for 'Dafydd' in Welsh. Gifford explained the science of the greenhouse effect to the analyst, who then asked if he would consider writing a brief for him. Gifford agreed. He saw it as an opportunity to influence the thinking of decision makers on an important subject.

'I wrote what I thought was a good version for a non-science audience on the greenhouse effect and some of its potential repercussions,' Gifford says.

Unfortunately, Gifford is not aware of any surviving copies or drafts as, when he retired, circumstance forced him to dump his papers. Attempting to recall what he told the agency 40 years earlier, Gifford says he delivered what 'would have' been a 'six-to-eight' page document that outlined 'the thrust of the grave risks down the line'; 'the long timeframes to go from recognising the problems to seeing it emerge'; and 'the long timeframes needed to start turning society around from emitting CO_2 from burning fossil fuel to some alternative energy mix.' It did not address the risks posed by methane emissions, as it was an emerging field of research and he was not familiar with it at the time.

Looking at the final version, now publicly available online, Gifford says he still recognises his writing. Much of the report reproduced his brief without change, but Gifford says the Welshman 'modified' parts of it – roughly 20 per cent – inserting material about what climate action meant for Australian coal exports and rewriting sections in the language of an ONA analyst 'for presentation to Malcolm Fraser'.

'Afterwards, [the Welshman] was very happy with my contribution. He thanked me very much and took me out for lunch at the Kythera Motel on Northbourne Avenue in Canberra,' Gifford says. 'We had a nice Chinese lunch.'

'That was the last I heard.'

The report was prescient in more ways than its concern over coal exports. Crucially, the analyst built on Gifford's contribution to accurately forecast growing domestic opposition to fossil fuel use and export:

> Many countries and business corporations are intent on expanding the use of coal as fuel, apparently without full knowledge of the carbon dioxide problem. This is particularly true in developing countries. So far there is no anti–fossil fuel lobby comparable to the anti-nuclear groups, although some environmental groups are beginning to express concern. Perhaps because the problem is merely the gradual increase of a non-poisonous substance which has always been present, public alarm will only be generated by manifest change or a threat of it, such as a rise in sea level. Nevertheless, increasing awareness of the problem could begin to generate an opposition to fossil fuels, encouraged by pro-nuclear lobbies and environmental groups, in this decade.

All they were waiting for was a spark to set it off.

James Hansen was way out on a limb. As NASA's leading climatologist, he was about to drop a bombshell on the American public – and by extension, the world. It was June 1988 as he sat before a senate subcommittee in a tan suit with a red tie, his hair combed to one side to hide the spots that were thinning.

Hansen started his presentation with three key statements. In quick succession, he told the packed hearing that the earth was warmer in 1988 than at any time since instrumental measurements began, that it was already possible to detect the effects of global warming, and that computer simulations showed how the belt of waste gases floating around the earth was beginning to amp up extreme weather events.

A videograph was displayed showing temperature changes over

the previous century. After working through the numbers for those listening, Hansen concluded by telling the committee that he could say with '99 per cent confidence' the world had warmed.

'Altogether the evidence that the earth is warming by an amount which is too large to be a chance fluctuation and the similarity of the warming to that expected from the greenhouse effect represents a very strong case, in my opinion, that the greenhouse effect has been detected, and it is changing our climate now,' he said.

What Hansen had given America's political leadership was a tangible truth, even if others in the scientific community felt his statements approached alarmism. Hansen's paper containing these measurements had yet to be published, and other, more cautious scientists – some of whom were possibly jealous of their colleague's new-found fame – voiced their objections in the ensuing debate. Natural systems were complex, they said. Evidence suggested the planet was warming on average, and there was a strong case supporting the theory that humanity had a role in it, but there were still unknowns, and others reviewing the same evidence could, they insisted, have reasonably come to a different conclusion. It wasn't yet feasible, they argued, to categorically exclude the possibility that the warming trend was caused by natural variability – and they suggested it probably wouldn't be for another decade.

This was also how the Australian media reported the issue at the time. *The Canberra Times* ran stories with headlines like 'Circumstantial case for the greenhouse effect' and 'The clouded view of the global greenhouse'. Yet Hansen was right. In the same year that he gave his testimony, the United Nations formed the International Panel on Climate Change (IPCC), with Swedish scientist Bert Bolin as its head. Bolin had been studying the greenhouse effect for three decades; however, concerned about the potential consequences of using alarmist language, he emphasised extreme caution in the body's reports, to the point where it might be said that the organisation underplayed the threat.

What Hansen could not have known was that the global oil industry had anticipated this very moment. The year before, Lee Roy Raymond, a man notoriously hostile to the notion that burning oil, coal and gas had any meaningful effect on the climate, was appointed to head Exxon, ushering in a new era of leadership. This changing of the guard brought with it a change in ideology. Under his watch, the company scrapped its research programs investigating the greenhouse effect and took a more muscular approach to the question of climate change.

In the same year that Raymond was elevated to head up the company, an IPIECA symposium was convened in Baltimore. Industry experts from across the world met to review the current state of knowledge about climate change and discuss what policies might be implemented to address the issue. Ostensibly this was about helping governments make better choices when the time came to act. The reality was they were war-gaming potential scenarios as an exercise in long-term risk management – the risk being that governments might actually move to address climate change. Drastic action seeking to rein in emissions by curtailing fossil fuel expansion had to be headed off early.

As Hansen's testimony was reported around the world, Raymond's Exxon was already mobilising. An internal memo written by the company's head of corporate research, Frank Sprow, warned that any global push for climate action would be bad for business.

> If a worldwide consensus emerges that action is needed to mitigate against Greenhouse gas effects, substantial negative impacts on Exxon could occur. Any additional R&D efforts within Corporate Research on Greenhouse should have two primary purposes: 1. Protect the value of our resources (oil, gas, coal). 2. Preserve Exxon's business options.

In a 2023 interview with *The Wall Street Journal*, the news outlet which went on to publish the memo, Sprow confirmed his suggestions had been adopted as company policy.

Meanwhile, as Sprow was outlining how the company might mount a product defence campaign, a group led by Exxon's manager for science and strategy development, Duane LeVine, met at Total Oil's headquarters in Paris. They called themselves the 'Working Group on Global Climate Change', and their role was to 'coordinate members' efforts to understand the issue, to promote their support for education and research, and to engage with international activities'. The focus of the group at that initial meeting would be to draw up what they called 'no regrets' response strategies that governments could adopt in response to any call to action. These were industry-friendly policies designed to appeal to governments nervous about the disagreement among the scientific community regarding Hansen's claims – governments who wanted to look like they were taking the issue seriously but didn't want to bet the house. 'No regrets' meant taking action against climate change – without doing anything that might harm the economy.

After all, wasn't it always better to be safe than sorry?

Even as the international oil industry was organising vertically through global organisations, it was also looking horizontally at other industry groups whose values were aligned. Over in the US, the Global Climate Coalition was founded as an industry alliance to aggressively lobby against climate action. Its founding members included all the usual suspects from the US oil industry, but also companies as varied as BHP, Dow Chemical, General Motors and IBM. Australia's equivalent, the Australian Industry Greenhouse Network (AIGN), would operate similarly to its international counterpart over the next decade, with APEA and AIP playing a key role. Until that point, however, organisations like APEA and AIP were either out on their own or operating in ad hoc partnerships as the need arose.

It worked like a one-two punch. AIP, with its international connections, went first. In 1989, the organisation published its policy position on climate change and hit the ground running. AIP's membership that year numbered 1698 individuals and 55 corporate members – including just about every major oil company operating in Australia – with branches in every state. The organisation presented this position at various seminars and provided copies to the Senate Standing Committee on Industry, Science and Technology, to state governments and to the federal opposition.

AIP's position paper and APEA's internal position would echo the language of similar documents published by the American Petroleum Institute and IPIECA during this period. Both deployed the rhetorical tactic of asking strategic questions to suggest controversy and feed doubt. Scientists did not yet 'fully understand all the mechanisms involved' in global warming, AIP claimed, warning policymakers that 'governments should be careful before adopting policies which carry a high economic cost and which may prove to be unnecessary or misdirected.'

With a wink and a nudge, the statement exploited the lack of understanding among the wider public about the scientific method. Scientists regularly debate or refine the finer points of a theory without undermining the theory itself, but this strategy left the science open to attack by bad-faith actors. At the same time, the industry preyed on the worst instincts of a political leadership that innately feared being trapped in a 'gotcha' – it would be a career-killing blunder to commit themselves to action on allegations that later proved wrong or inaccurate. By this time, there was a general awareness among members of the global petroleum industry that the science was more settled than they were willing to let on. Exxon, BP and Shell had their own climate change programs, but they were quite happy to invite governments to gamble on the possibility the science was unclear. For good measure, AIP's position paper informed its reader that 'the greenhouse issue is a global one', calling for 'international co-operation and co-ordinated policies'.

'Unilateral action by Australia to limit greenhouse gas emissions is not appropriate as it has minimal impact and would service only to penalise local industry and living standards by imposing costs on them not borne elsewhere,' it said.

As a strategy, it was clever, but it didn't stand up to basic reasoning. If pumping carbon into the atmosphere could be compared to pissing in a swimming pool, fixing the problem didn't mean pointing to the next person over and saying, 'I'll stop when they stop.' It could even be argued it wasn't entirely necessary to convene a meeting of the parties to discuss non-binding solutions before any action could be taken. People just had to stop pissing in the pool. Having done so, those who had stopped would have the moral authority to yell at those who hadn't, until they finally did.

Reason, however, wasn't the point. What mattered was repetition. As bullshit, the ideas were immune to refutation, and because they were useful to a particular constituency, they stuck. The suggestion that Australia was a small player globally wasn't accurate when the impact of the country's fossil fuel exports was considered relative to its population size – yet the idea that it was utterly powerless to break from the herd would become an article of faith in certain policy circles for decades. It would become a frequent source of friction when those in government or the public service finally sought to do something about the issue.

These AIP talking points would also be echoed by APEA, which, in its annual report that year, disclosed taking a 'major interest' in greenhouse policy. In what is currently the earliest known published statement by the association on the issue, it explained that the organisation was 'prepared to be involved in responsible and effective policy-making'. Though APEA understood that 'to do nothing at all' about climate change was not 'an acceptable option', it stressed that 'far more research' needed to be done into the impact of the greenhouse effect and advised that any major energy policy changes would be 'premature and unwise'.

Specifically, it argued any policy action in Australia should not involve:

- Costly unilateral actions which will lower Australia's living standards relative to the rest of the world, e.g. The ongoing search for and production of petroleum energy sources in Australia should not be discouraged as increased crude oil imports will have adverse macroeconomic consequences through further deterioration in the balance of payments and present greater environmental problems due to increased sulphur content and heavier nature of imported crudes;
- Commitment to artificial reduction in 'Greenhouse' gas

emissions which mirror global targets – there are environmental benefits associated with burning Australian fossil fuels compared with overseas fuels;

- Significant interference with market forces – activity directed at large-scale reductions of 'Greenhouse' gas emissions can be commercially based on a global scale.

The association also recorded how it emphasised in its discussions with political representatives that the greenhouse effect was 'a global issue, not a national or local one'. 'Therefore, its resolution must be global.'

The irony is that by 1992 the IPCC's work showed James Hansen had been right all along. Much later, in 1996, its reports provided clear proof of the impact of burning fossil fuels on the composition of the atmosphere, as demanded by its critics, who insisted that more evidence was needed before any action could be taken. However, when they provided this material, the scientists who had done the work would be attacked by industry and by the IPCC itself.

The AIP and APEA reports signalled a profound shift in attitude and outlook for the oil and gas industry. Back in 1973, they had been openly discussing the risks posed by climate change and what might be done about it. By 1989, they were taking part in a global campaign to spread misinformation, emphasise doubt and generate pressure to stall or slow attempts by government to act. The industry strategy was clear. First, they fought to ensure individual governments did not act alone on climate change and then, having successfully helped make the response to climate change a global collective action problem, their representatives later worked to frustrate this process with their friends and allies.

The words 'no regrets' would define Australia's response to climate change for a decade and haunt it for two more. This policy would land in Australia in the flurry of activity that took place around 1989. It is not clear how these words made the jump from the Paris offices of Total Oil to Australia, or when they became assimilated into Australia's bureaucratic thinking, but a Hawke government press release suggests it was sometime around October 1990.

As the decade came to an end, environment minister Graham

Richardson and his successor Ros Kelly were pushing for deep cuts to the country's emissions. Australian scientists had recently attended an international climate summit in Toronto, bringing back with them its call for the country to stabilise its emissions at 1988 levels, and to reduce its emissions by 20 per cent by 2005. When considering whether to implement these targets, the Hawke government wavered. Thanks to an intervention by then-treasurer Paul Keating, cabinet resolved only to adopt policy on a 'no regrets' basis, a position made clear in a subsequent press release:

> While recognising the need to restrict emissions and to aim for a 20% reduction, the Government will not proceed with measures which have net adverse economic impacts nationally, or on Australia's trade competitiveness in the absence of similar action by major greenhouse gas producing countries.

In other words, Australia would do nothing that might harm its oil, coal and gas exports.

Having helped patch this thinking into the Australian political programming, Orchison wouldn't stick around at APEA during this next phase. He had impeccable timing as a lobbyist and a wizard-like ability to summon deputy premiers from the ether before stuffing their mouths with his words. These traits had helped bring APEA back from the brink of the Whitlam years, but it was still an organisation with a narrow vision. When Orchison left to head up the Electricity Supply Association of Australia – which represented power producers and would become a crucial plank in a broader industry campaign against climate action – his hand-picked successor, Richard 'Dick' Wells, would take over and set about changing everything.

If Keith Orchison was a lobbyist, Dick Wells was a creature of government. Where Orchison was flashy, quick-witted and slick, Wells was roughly 85 kilograms of dull packed into a trim grey suit – a man perfectly bred for the Keating era of the 1990s.

Liberalised financial markets and public-service reforms had ushered in a new pragmatism – politics was no longer political. Evidence and consensus were now cardinal virtues and tone mattered almost more than content. Where once politics was a tussle between interests, the measure of success now was the capacity among 'serious people' to make compromise. Surprises were to be minimised and public displays of conflict avoided. The only true failure was to lose a seat at the table by saying or doing something outrageous.

Wells was an 'insider's insider', someone with a good reputation in government who intimately understood Canberra's plumbing. He had graduated in science from the Australian National University before joining the public service and landing in the Department of Resources and Energy. There he headed up a team researching the impact of the Ranger uranium mine on Kakadu National Park and oversaw the former test sites of Woomera in South Australia and the Montebello Islands in Western Australia. Wells had also been a senior adviser to Labor ministers Peter Walsh, Gareth Evans and would later advise John Kerin. When he wasn't employed by government or the public service, he had done a stint with ACIL Australia, the oil and gas industry's preferred consultancy.

Those who encountered Wells during this time mostly remember him for being nice. Upon taking over APEA, the amiable Wells lent the organisation a veneer of seriousness and respectability. If Orchison's approach was art, Dick Wells' was science. Like Orchison, Wells was no oilman – which would prove his biggest asset. During his tenure, Wells' task was to secure a clean break from the past. He would remake the organisation as a professional outfit and give it a clear structure so it could more systematically pursue its goals. Wells may not have been one of them, but because he worked for them, Max Roberts was dispatched to advise him in the early days. Roberts was a loyal industry man who had made his career at Castrol before being headhunted by Burmah Oil – if you cut Roberts, people joked, he bled oil. He was also a walking Rolodex of the Australian business community, and his job was to secure Wells' entrée into the fold by setting up a series of lunches and dinners.

To facilitate all the necessary politicking this required, APEA voted in 1994 to move its office to Canberra. The decision would allow Wells to 'constantly walk the corridors in Canberra', in the words of one former APEA executive. When the decision was made, all but one of the association's staff quit to remain in Sydney, leaving a blank slate for Wells to remake the organisation. He responded by aggressively recruiting a crack team from within the public service and government – a strategy the association pursues to this day. Among the new hires were Lynda Gordon, who worked in public affairs for the New South Wales Chamber of Mines; Peter Cochrane, a former adviser to Simon Crean; Tony Noon, former acting assistant director of the Queensland Department of Minerals and Energy Geological Survey division; Noel Mullen from the federal Department of Resources and Energy, and Cath Sutton, an accountant from Queensland who worked with the Real Estate Institute of Australia. Niegel Grazia, a former adviser to the WA government and assistant director of the Pilbara Development Commission, was brought in a year later to manage the organisation's Perth office when it was opened in 1995.

This kind of private-sector experience was considered valuable among ambitious policy wonks of the mid-1990s. The idea was that apolitical, technocratic experts with experience in both industry and government who were fluent in bureaucratic lingo were a net positive. Going private was attractive to ambitious professionals hoping to pad out a résumé. Their ability to translate – to act as liaison and smooth over misunderstandings – theoretically led to better decision-making over the long haul. The noble ideal was that the benefits flowed to the government, but experience shows it only ever worked in reverse. The oil and gas producers hadn't just bought the contact books of these former bureaucrats. These new employees would also serve as 'trip wires' to give advance warning of change within government and the public service. Moreover, decades of corporate memory had been transferred into private hands. When no-one remembers the past, those writing policy for the future can't know they are fighting the same old battles, the same way. Every confrontation becomes an ambush.

In 1996, the organisation officially added an extra 'p' to its name, changing from **APEA** to **APPEA** – the Australian Petroleum *Producers* and Exploration Association – and commissioned a new logo. It was a rebadging, intended to signal the start of a new era with a new staff, a new office and a new sensibility. Atop the nerdy irascibility of an industry association composed of engineers, a slick, corporate façade had been built that aimed for professional respectability.

When he was done, Wells would be rewarded for his work. He had so impressed the resources sector with APPEA's makeover that he would be offered a position heading up the Minerals Council of Australia. The organisation had recently gone through its own Enron-style rebranding after its single-minded hostility towards Indigenous land rights soured its relationship with the Keating government. Wells' appointment to head up the organisation, which represented the bulk of Australia's coal producers, cemented APPEA's reputation as a proving ground for future industry leaders.

And to be fair, what Wells had achieved *was* objectively impressive. With his team, he had crafted a finely honed mechanism through which the nation's oil and gas producers could intersect with government, so they might better insert themselves into the legislative process. In effect, Wells had built the corporate equivalent of a Glock and left it loaded on the table, ready for his successor.

8

Let Them See We Are Not Ogres

It wasn't until Lucy Snelling, head of corporate and commercial at Queensland gas company State Gas, appeared on a panel at the 2022 APPEA conference in Brisbane that I really began paying attention as she described what was being taught about the oil and gas industry in Australian schools as 'brainwashing'. Until that moment, I had been sitting ringside in the main theatre, only half-listening as the industry panel of heavy hitters ran through all the usual talking points expected at an event of this nature. Alongside Snelling sat Chevron's Kory Judd, Woodside's Meg O'Neill and ExxonMobil's Dylan Pugh. They were about 45 minutes in when they started taking questions from the audience, and it was the third question, or maybe fourth, when it happened. Panel host and TV journalist Ali Moore read out a message from an audience member that blithely asked, 'What's our messaging to the young generation to attract talent to keep our industry going?'

The first thing that stood out was the phrasing. Whoever asked it – questions were submitted through the conference app – didn't want to know how oil and gas companies thought they could repair the trust lost by three decades of playing for time on climate action. They didn't want to understand why kids might not be keen on an industry whose employees had a higher than average cancer risk and whose primary product was responsible for driving climate change. What they were looking for was the sales pitch, that special sequence of magic words that would convince the next generation to sign on, guilt-free.

But it was Snelling's candid response to the question that seemed to offer a peek behind the curtain, a hint at the conversations that were being had away from public view. The other members of the panel, practised diplomats all, had responded with easy platitudes about how oil and gas companies needed to give young people a chance to be 'part of the solution', but Snelling's answer was personal. She had two sons studying to become engineers, she said. Both had told her that fossil fuels were a 'dying industry'. It was clear, Snelling explained, that oil and gas companies had to do more to get into schools and correct this perception at the source.

'So, I think it's really a huge challenge, some of these issues about public perception about the industry, and I think we have to work very hard,' she said. 'I think that there's a big task in the schools – we have to get into the schools and try to work with the teachers, because there is this widespread perception, I think, amongst teachers and staff, that these industries, that the world is moving away and there's no future.'

Like everyone else on the stage that day, Snelling was sincere in her belief that the world needed oil and gas. It wasn't possible, in her view, to build out hydrogen, batteries or extract critical minerals without the skills developed over 60 years of oil and gas operations in Australia. Turning back to the question of recruiting and retaining the next generation of petroleum engineers, she laid out a two-step process: get the kids into training programs, and then let self-interest take care of the rest.

'Once they get the skill base, they'll go where the jobs are or they'll go where the money … where the opportunities are,' she said. 'But I think we've got to try and sell it on a basis that, for these people, because some of them are well past the stage where you can fight that brainwashing that they've received.'

There was the briefest of pauses as Snelling caught herself using the word 'brainwashing'.

'Perhaps I shouldn't use those terms,' she said, as the audience laughed.

Moore, being a journalist, immediately followed up, asking whether she actually meant what she said.

'Yeah, look, it's a lack of education, but it's a real challenge,' Snelling said. 'And I think it comes back to the trust question – trust issue – as well. It's like the old thing about lawyers: lawyers are terrible, except my lawyer.'

Months later, I wrote a feature story about fossil fuels in schools and as part of my research I followed up with Snelling to clarify her position.

'I very much regret using the term "brainwashing" in my comments at APPEA. The term is highly emotive and negative and does nothing to help debate,' she said. In this case, she said, her word choice had detracted from the bigger point she was trying to make.

'The challenge, as I see it, is that the picture of the role of oil and gas in climate change and our future tends to be simplistic: *It causes climate change; the industry must stop,*' she said. 'This is, I think, a widely held view. However, our world is rarely so simple, and it is not so here.'

Snelling explained that she did believe emissions from oil and gas contributed to climate change, but she also wanted to make it clear she took issue with the idea that the industry was 'like a tap that can be just turned off'.

I knew, having hung around these conferences, that this was a common industry talking point, and that it mischaracterised the managed phase-out that critics of the industry were actually calling for. When the climate movement said 'end oil', what it wanted was a carefully managed plan to rapidly end demand as the industry wound down over time. Industry people tended to treat this suggestion as if people wanted to throw them out of their jobs tomorrow and the world into chaos. Whether this was a sincere belief among industry people or merely kayfabe – that is, the deliberate performance of misunderstanding in public debate – it was in their interest to allow such a bad-faith reading to percolate.

Snelling's primary concern, she added, was to 'minimise harm' during any transition away from oil and gas. Part of the problem, she said, was that the 'workforce in the oil and gas industry (and mining generally) is ageing, and not enough young people are joining'.

'So, an education which leaves young people thinking that there is no future in this industry, that does not see the skills the industry develops as relevant and necessary for the future, is highly problematic,' she said.

To be fair to Snelling, she wasn't the only one saying things along these lines at the 2022 APPEA conference. Considerable airtime had been given to debating how industry might recruit the next generation. Another presentation had offered suggestions about how human resources departments could recruit and retain millennials; a separate panel featured Michael Holmstrom, co-founder and CEO of STEM Punks, a third-party education provider, who explained how he broached tetchy subjects like climate change in the classroom.

Holmstrom was an interesting figure, in that he was not from the oil industry. He spoke with the bravado of a Silicon Valley founder, and his company ran workshops in schools for kids so they could learn about technology. His organisation had partnered with Santos that year to run an education workshop for high school students at the conference.

'I think there's also an issue of engaging with kids too late,' Holmstrom explained during question time at the end of the session. 'When you get to middle school, it's too late already – kids have made up their mind. It's sort of late primary school, early high school, where kids form a structured opinion based on whatever information they get.'

The consensus seemed to be that if petroleum companies really wanted to get the next generation on their payroll, they had better get them young. Here was a crop of oil and gas industry figures, and their contractors, talking openly about how to better get into schools to shape young minds.

Leaving the conference, I wanted to know more. As I began to learn about how industry had long since set its sights on the nation's schools, I found something unexpected: an organised public relations campaign by the Australian petroleum sector to convince the next generation to love oil and gas.

As a veteran high school teacher studying a graduate certificate, Elaine Grossman appeared decidedly out of place when she began her presentation to the oil industry at the 1994 APEA conference. The industry was only just beginning to organise itself. A few years earlier it had coordinated with its global counterparts in an effort to stop binding action on climate change, but now it was laying the foundation for what would become a domestic crusade to stop John Howard signing the Kyoto Protocol. Those were bold moves on the chessboard, actions intended to bend policy to industry's will. They were also quick wins. Those who took a longer-term view understood that it wasn't enough to impose their will on decision-makers. Change – real, lasting change – had to be socially embedded within the communities they operated in. Achieving this outcome required smaller, quieter interventions that might not mean much in isolation but over time added up to something significant. Schools were a natural focal point for an industry increasingly concerned about growing hostility among the public – as Grossman herself would explain: 'Local schools are ideal places through which industry can become better known.'

In one sense, Grossman was articulating a line of thinking that was all too familiar within industry. For as long as resources companies have operated in Australia, they have sought to link themselves to educational institutions to secure the skilled labour they needed for their operations. Back in 1949, for instance, Australian mining companies chipped in to establish a chair of economic and mining geology at the University of Adelaide, beginning a long association between industry and university's geology department.

As a distinct group within the broader resources sector, Australia's oil industry began to collectively think about how it might infiltrate higher learning around the same time petroleum engineering degrees began to proliferate across the country. In 1966, APEA formed an internal education committee to 'provide a link between universities and the requirements of the exploration industry'. By 1981, the federal government established dedicated training centres for petroleum engineers as part of a broader push to address a skills shortage, at which point the oil industry's focus began to shift from universities to high schools. With the government encouraging tertiary-level training, industry directed its energy to finding creative ways to funnel high school students towards these programs. To help smooth the way, APEA sent out school information kits and slide presentations to 1800 institutions around the country.

Meanwhile, individual companies were busy running their own ad hoc partnerships, education and outreach programs. Usually these focused on the local schools clustered around their industrial operations, often arranging guest speakers or site visits to refineries. Most were activities run by individual companies in isolation from each other, each delivered with varying levels of skill or enthusiasm. Some companies considered it their obligation as a major employer and looked at school programs as a way to give back to the local community. Others had little interest in anything that didn't generate an instant profit and so only made a token effort at engagement.

Grossman proposed taking these initiatives a step further by unifying them. Where, in the past, efforts by individual companies had been isolated, she proposed binding them together, to 'supplement' their programs through the efforts of industry associations.

It was, objectively, an ambitious pitch from a woman who understood the education system but only had limited experience with the industry. According to her official biography, she had worked as a biology and environmental science teacher, serving 15 years in Victorian public schools, before she was seconded to the Australian Institute of Petroleum in 1993 as part of a teacher-release-to-industry program coordinated by Deakin University. Grossman's biography raised more questions than it answered, like what motivated her shift, and how she felt about her work with the benefit of hindsight, but when I tried to track her down by seeking out former colleagues and people who shared a surname – no-one knew what happened to her.

Thanks to her professional experience, Grossman brought to the program a certain finesse and a focus on young children. APEA created the Schools Information Program in partnership with the Petroleum Club of Western Australia in 1991 with some success. The AIP–APEA partnership, however, was more ambitious. Grossman was already a year into her role when she briefed the industry at the APEA conference on her work. In her presentation, she laid it all out for the executives, many of whom, until that point, were unconvinced about the need to get involved in education.

'The petroleum industry suffers from a poor public image and is not treated objectively in education circles,' Grossman said. 'The reasons for this stem more from ignorance than a deliberate ploy, and for this reason it is important that efforts be made to improve the industry's position.'

The public hatred for oil and gas executives was no secret. The gaping hole in the ozone above the Antarctic had raised public awareness of the damage corporate activity could do to the environment, as did studies showing that leaded petrol caused permanent brain damage. Periodic spikes in the fuel price didn't help, either. Then, in 1989, the *Exxon Valdez* supertanker ran aground off Alaska. The legacy of that spill seeped deep into the 1990s as oil executives became fair game in popular culture. Children's TV shows like *Captain Planet* and *FernGully* explicitly presented extractive industries as evil to a generation of young millennials. This type of cultural product only fed industry concerns that hostility to fossil fuel producers might trickle down to a new generation.

Whether she realised it or not, Grossman's vision sought to put a handbrake on a cultural shift. During her presentation, she made clear that the schools program was an exercise in public relations, its primary goal to 'improve the public perception of the industry and to affect the way in which it is portrayed'.

'By assisting the wider understanding of petroleum matters, the industry will enhance its own viability and ultimately its long-term survival,' she said. 'An involvement in education can create familiarity and reach the people who have a stake in the industry's future.'

To achieve this, Grossman offered a plan. The AIP–APEA partnership would make its own classroom resources and curriculum materials, connect schools with individual companies, arrange site tours for both teachers and students, organise industry guest speakers, liaise with teacher associations, provide professional development for teachers and negotiate sponsorship arrangements. To smooth over fraught internal industry politics, Grossman also stressed that any efforts by the AIP–APEA partnership would not 'supplant' individual programs already being run by companies but rather 'support' them.

'A correct approach to schools is vital,' she said. 'While a few teachers may be hostile to the petroleum industry, most are not. The main reason for misinformation about the industry is simply that teachers are ill-informed about it. Before teachers can teach more accurately about the industry, they have to understand it.'

By the time she was finished, Grossman had outlined the creation of a sophisticated influence campaign targeting Australian schools on a bigger, more organised scale than had been attempted before. Just as the Commonwealth Bank had secured a generation of customers through the Dollarmites program, Grossman had shown the petroleum producers how they could go about building their own loyalty program. She told them exactly which buttons to push – how to make sure school materials were 'curriculum aligned', how to target teachers with specialised training days, and the importance of offering them ready-made lesson plans that made their lives easier.

Where the industry's previous efforts were isolated, reactive and technical, Grossman brought them together in the culmination of decades of advocacy. As Peter Power, former APEA chair and director of Ampolex, explained in a book recounting the association's history, a one-off school tour to the Australian Marine Oil Spill Centre was useful, but the real prize was to have a physical presence inside classrooms.

'To go into a classroom and speak to the kids about the industry and let them see that we were not ogres was important,' he said. 'We showed we had a human face and we were producing a basic resource for everyone's benefit.'

★

Tracking the oil and gas companies' efforts to get into schools turned out to be as simple as visiting the library. I'd found the first evidence – a workbook for high school students – searching the University of Adelaide library collection. I pulled another children's book from a stack of volumes I'd requested from the National Library in Canberra. A poster and set of slides were still housed in the State Library of New South Wales.

These were early relics of Grossman's schools program – which hadn't lasted long. Once the first computers were installed in schools, everything soon shifted online. The AIP–APEA program immediately moved to take advantage, printing CDs with educational programs for high school students and producing a fully animated story of oil and gas called *Petromania*. With enough time, dedicated websites were created to host these materials, allowing teachers to download them anonymously and on demand – and it was at that point efforts to coordinate industry-wide education efforts fractured. The AIP–APEA education program didn't survive the internet, and educational activities were mostly left up to individual companies once again.

Those efforts, however, remain ongoing – and some of them have become influential. Today, a website maintained by the Woodside Australian Science Project, a partnership between oil giant Woodside and Earth Sciences Western Australia, remains hugely popular among teachers who consider earth and space sciences dull and difficult to teach. Though many of the materials hosted there are uncontroversial, it offers worksheets for Year 10 students on the 'greenhouse effect' that suggest humanity's role in driving climate change remains 'a point of some debate'. Another section suggests students should be 'grateful' for greenhouse gases in our atmosphere because they 'keep the surface of this planet warm enough to support life'. None appears to have been updated since they were first published in 2010.

The earlier materials remain a testament to the industry's ambition. Flicking through them, a clear pattern emerges. Each seems to follow a similar formula: they prime their impressionable readers with some general background information, present a problem, introduce a suite of solutions and then heavily point towards the industry's preferred outcome. At no point are readers allowed to imagine a world post-oil.

One example is *The Australian Petroleum Industry Resource Book*, a booklet published for high school students in 1993. On the weathered cover, a silhouette of an offshore gas well stands against a fading sun. The sky is coloured with the hues of twilight: soft yellows bleeding into blue, purple and orange. Bobbing on the sea is a boat moored to the well's platform, a helicopter buzzing overhead.

Looking at it more closely, the booklet employs a clear bait-and-switch. It does provide genuine geological and engineering information about the process of finding and refining oil, but it also opens with a lecture on economics. To clear up any concerns its reader may have about 'peak oil', the booklet cites works by Texas A&M University professor S Charles Maurice and Max Singer as authorities. It did not disclose that Maurice was an academic consultant for the tobacco industry, nor that Singer was a co-founder of the Hudson Institute, a free market think tank that received funding from ExxonMobil and the Koch family foundation as part of a broader campaign to spread disinformation about climate change.

In fairness, it would have been hard to find this out. Thanks to the internet, it now takes all of 30 seconds to check, but a high school student in 1993 was not so lucky. Neither was the booklet's author, Peter McGregor, who was pictured on the back cover. When I contacted him to ask about it, the 82-year-old could not recall ever having worked on it. He had been a third-party education consultant for big corporates but had long since retired, had no memory of even being involved and was surprised to see his younger self on the back cover. He had once worked a job for Shell, he said, which, he speculated, was how he might have landed the gig.

McGregor was happy to tell me what he did remember, though, and described his general process for selling his booklets to schools.

'What I would do is I would send a book out, addressed to the science teacher or the geography teacher, whoever I thought was most appropriate, and ask if they wanted any more to get back to me,' he said. 'And if they wanted a class set of twenty-five, I'd send them some.'

I didn't press McGregor on what the book said – it seemed pointless, as he had no recall of his role in producing the booklet. It seemed likely he had been bamboozled by the same misinformation campaigns that made Maurice and Singer prominent.

Unable to find out any more about the booklet's creation, I took a closer look at its contents. Having primed its readers with a brief discussion about peak oil and a basic lecture on free market economics, the booklet directly addressed environmental issues in a section titled 'Environmental Impact'. Two long introductory paragraphs directly linked rising living standards to oil extraction and presented environmental harms as a necessary by-product of progress. When it finally addressed climate change, it explained that 'many scientists' had 'major' concerns about 'the possibility that burning fossil fuels may warm the atmosphere and affect the cosy greenhouse in which we all live', but it also stressed that 'the exact facts are uncertain' before assuring its readers that governments were 'taking action'.

Several countries were taking steps, it said, to 'start reducing greenhouse gas emissions in the hope that a future catastrophe might be avoided', but it also sought to cast those who raised concerns about climate change in Australia as rich, decadent and out of touch:

> As a result of increased awareness, there is concern within the community that the environment should be given priority. This is especially true in rich countries like Australia where life is fairly comfortable and secure for most people. Some environmentalists and others speak of 'saving the planet'. They question many of the activities upon which life as we know it today depends.

It then declared: 'Looking a long way ahead, cleaner energy sources may be asked to do more and more of the work that petroleum does today.' But if this was a show at even-handedness, it also reminded its reader that 'most forms of renewable energy are more expensive than petroleum products,' and that they were unreliable or unavailable at scale. It then concluded by answering its own rhetorical question.

'These energy sources may be used to solve the problems of future generations. But ours?' it said. 'A reduction in the demand for oil seems unlikely in the years ahead. Apart from recession, or attempts by governments to encourage us to consume less (mainly by increasing the price), the need for petroleum products will continue.'

In other words: petroleum was forever – for now.

Another example I found of this kind of industry propaganda was *The Big Book of Oil & Gas*, published a few years later, in 1997. This one was a picture book produced by AIP and APEA, illustrating the scale of the industry's designs on the next generation. The big, silly cover was reminiscent of a children's colouring book. Off in the background, a distant purple volcano was erupting. A happy red stegosaurus stood on a beach in the middle ground and a smiling green diplosaurus munched on leaves in the foreground. Beneath them, buried deep down in the soil, were bones of critters, which would later become oil.

The Big Book of Oil & Gas was full of colourful pictures of dinosaurs, smiling kangaroos set against oil wells and photos of seismic trucks used to search for crude. A map on the inside cover showed the position of every pipeline, refinery and site of production in Australia at the time. One section helpfully informed its young readers that 'natural gas is friendly to the environment and is very efficient when used correctly.' Turning to the back, I found a note instructing teachers on how to deliver the material in a classroom. It advised:

> Prior to the introduction of the big book, teachers should conduct tuning-in activities to familiarise young people with the concept of oil and gas and stimulate the idea that petroleum is more than the product which is available at service stations.

This was a common classroom technique for young children – but in this case the teaching materials were propaganda created by the oil and gas industry.

Going back to the start, I looked more closely at the information on the first few pages. Certain keywords were in bold for emphasis. I began to read them out loud to myself, as if I were in a classroom.

'Everywhere we look, we can see the importance of **oil** and **gas** in our lives,' it said. 'Large trucks take **petroleum** products from the **oil refinery** and deliver it to service stations around the country.'

I imagined a group of schoolchildren sitting around, cross-legged on the floor, looking up at me as I read. I thought about them repeating my words as I spoke. It was then I stopped and closed the book. If brainwashing didn't fit, the word 'indoctrination' came to mind.

Australia's oil and gas producers were never shy about their intention to get into schools – Grossman went so far as to say companies should be 'involved in the decision-making process at all levels to ensure that educational outcomes meet industry needs and expectations'. But even with this stated goal, there remained a problem: they still had to be invited in.

Teachers as a community were a strident bunch, and marching up to a school reception desk wearing a polo shirt emblazoned with an oil company logo wasn't likely to be met with a smile. The quickest and most efficient way around this issue might have been to help write the curriculum, but there would be no national curriculum until 2010. Each state and territory wrote its own, with different authors working on different subject areas and for different age groups. Even if a particular minister was sympathetic or could be leaned upon, there was no guarantee it would translate to outcomes. Individual schools often put their own spin on things, too, and for better or worse, teachers were ultimately free to deliver the material however they chose.

Some states have historically been more willing to cooperate with industry than others. Queensland, for example, allows industry direct access to high schools through its 'Gateway to Industry Schools' program. Its purpose is to funnel students from participating high schools into one of eleven industries by introducing them to specific fields through partnership arrangements. Though it includes companies operating in fields such as aerospace, film production and IT, the two most active industry groups are construction and resources – which in Queensland mostly means coal and gas producers. The partnership agreements between schools and individual companies are overseen by the Queensland Minerals and Energy Academy, an entity that was itself a partnership between the state government and the Queensland Resources Council. According to its website, the organisation 'provides a talent pipeline of employees into the resources sector'. Though the Northern Territory was said to be considering a similar approach, Queensland was the only state to run such a program as of 2024 – a reality which underscores the problem for industry.

If the companies couldn't get in through the front door, and they couldn't help shape the curriculum, what they needed was someone to smuggle them in through the back. The solution they landed on was the strategic partnership. At a basic level, it worked like a fake ID at a club. Oil, gas and coal companies, or their industry associations, would strike up a deal with a third-party organisation or institution. Whether it was a third-party education provider running workshops in schools, or external institutions like museums or science and technology centres, the companies would stump up some cash; in return, their brands were carried into places and spaces where they would otherwise be shunned.

Over the last two decades, there have been several examples, sometimes with cringe-worthy results. In 2014, the Dalrymple Bay Coal Terminal in North Queensland introduced a human-sized lump of coal named Hector – complete with picture book and television segment – as a way to appeal to kids, and was ruthlessly mocked for its efforts. In Western Australia, a Woodside-sponsored workshop run during National Science Week in August 2021 met with a similar response after it tried to teach Year 3 students how to drill for oil by having them suck Vegemite from sandwiches through straws. Questacon, the beloved national science and technology centre in Canberra, maintained a partnership with Shell and Japanese petroleum company Inpex until October 2022, when the arrangement reached its 'natural conclusion'. Though the centre claimed the companies had no influence over its exhibitions or publications, it uploaded videos to its social media promoting gas as a 'transition fuel' – a common industry talking point.

For all the money and time invested in these programs, the results were often mixed. A field trip got kids out of the classroom, and visits from Hector might have injected some novelty into their day, but whether the students actually took anything on board was another question. To engage them, it took something more: it took an organisation like STEM Punks.

Compared to other third-party education providers, STEM Punks stands out for actually being fun. As of 2024, the company has three headline programs: a sports innovation program, a space camp, and a sustainability camp. Its social media is awash with photos of happy kids playing with futuristic gadgets that have captured their imaginations. Looking at them, it is easy to feel a sense of envy. Even if the organisation talks about technology in the lingo of Silicon Valley, these are exactly the kinds of activities I would have loved as a kid.

On stage at the APPEA conference in 2022, STEM Punks' founder, Michael Holmstrom, cut an interesting figure among the oil and gas executives. He was both charismatic and, unlike others on the panel, not from the petroleum sector. Holmstrom had, in fact, started his working life as an electrical engineer designing radar systems used to monitor mine walls, before eventually spinning off his own company. The idea for STEM Punks came when his kids started at a school where they weren't teaching innovation in the way he thought they needed to.

During the panel's discussion, he lamented how kids were being 'bombarded with information' about the oil industry specifically, and mining more broadly. He explained his belief that there was 'a lot of misinformation and perception we've got to break down' and how, in his view, industry was 'engaging with kids too late'. By the time they hit high school, they had already made up their minds about the big issues, he said. When asked how he approached a subject like climate change in the classroom, he said it was necessary to address the issue head on, with 'honesty and transparency'. As an example, he explained how STEM Punks had organised a 'massive panel discussion' with high school students in Brisbane a couple of years back that sought to be 'non-partisan' and 'unbiased'.

'It was led by honesty, transparency, and a feeling of "yeah, we're not doing everything right, but hey, we're here, we're part of the solution,"' Holmstrom said.

What could have been a 'polarised discussion about us and them, yes and no' had become a 'really good dialogue' that he felt was good for students and for the industry. He did not clarify what he meant by 'non-partisan' or 'unbiased'.

Listening to this, I had several questions, but the biggest related to the apparent contradiction in logic. Holmstrom had called for honesty and transparency when confronting tough issues in the classroom, even as he seemed to be laying out what could, on balance, be reasonably described as a method for delivering industry propaganda directly to students at school. I wanted to understand how he squared these two intersecting circles.

So, I called him.

From the get-go, Holmstrom wanted to set the record straight. First and foremost, he was no climate denier – though I'd never thought he was. STEM Punks, he stressed, was an organisation led by the science – though I had never suggested otherwise. He also made it clear that he supported action to address climate change – though it had never occurred to me he might not. Mostly, he wanted to make clear that STEM Punks was no Trojan horse for the oil and gas sector to deny climate change – and I believed him. I had called him wanting to know what the deal with Santos was about.

'When you start a business, the first thing you need, you try to generate revenue,' Holmstrom said. 'We had a fair bit of revenue [from] the resource industry to get the business going and get it started. Today the work [we do for industry] is very minimal.'

A few years after STEM Punks was founded, he explained, Santos approached the company asking if they would be interested in a partnership. Together they began working to deliver educational workshops in regional areas. Through this association, the Santos logo landed on the STEM Punks website, where the oil company was identified as a 'key partner'. A video posted to the education provider's social media accounts showed kids at Roma State College in Queensland posing against a Santos banner. The equipment students worked with was stamped with both the Santos and STEM Punks corporate logos.

Now, Holmstrom said, things were different. A couple of years and a pandemic later, the company had grown, and the arrangement with Santos now brought in 'less than 1 per cent' of STEM Punks' total revenue – though he would not say how much they actually received. The programs they delivered with Santos were small relative to STEM Punks' overall output, he stressed. Asked about the text on the website describing Santos as a 'key partner', Holmstrom said it was out of date, not correct and needed to be removed. Regarding the video on the website showing smiling children standing in front of a Santos logo, Holmstrom said his company's workshops were 'in no way' about promoting fossil fuels. Rather, they were about 'empowering kids to learn critical thinking skills' and to 'learn skills to be part of the solution' to climate change.

I couldn't blame Holmstrom for the impulse to scrub all mention of his company's relationship to Santos from his website, but leaving it up at least acknowledged its existence, allowing schools and parents to make an informed choice about whether to work with STEM Punks – and there were still plenty of reasons why they might. Removing this information would mean schools had no way of knowing they were hiring an education company that, as Holmstrom confirmed, still received money from a fossil fuel company.

It was around this point in the conversation that we came to the heart of the issue. Santos was a petroleum company whose products were contributing to the destabilisation of the climate – an existential threat that future generations would need to grapple with. I wanted to know whether he thought the association with

an oil company like Santos might taint the other good work STEM
Punks was doing elsewhere.

What followed was an odd exchange. Instead of answering my
question directly, Holmstrom emphasised again that Santos was
'such a smaller player' and the work they did together represented
only a fraction of STEM Punks' activity. If that were the case, I
asked whether he would consider ending the relationship, owing to
Santos' role in contributing to climate change.

Holmstrom answered that he was 'not actively looking for a
replacement' – though this was not my question. At another point
in our conversation, he outlined his belief that there was value in
taking a 'collaborative approach', saying, 'Rather than sitting on the
sidelines, I'd rather work with industry.'

'I believe the fossil fuel industry knows there is change needed,'
he said. 'They know the population wants change in moving away
from fossil fuel measures to sustainable measures.'

It felt almost like we were speaking two different languages.

Around this point I started probing for some kind of ethical
baseline. I found it when I asked whether he would consider working
with a tobacco company.

'We wouldn't,' he said. 'It's not right. There is a fundamental
culture in what we do. We want to empower kids to be part of the
solution.'

Having established that he did have concrete ethical boundaries,
I began to work outwards to find out how he would apply them in
similar situations. The more I probed, the more reluctant Holmstrom
seemed to clearly rule out working with any other companies or
industries. It was only when I pushed that he confirmed STEM
Punks had previously turned down two partnership offers. The first
was from a gambling company and the other he would not disclose.
As STEM Punks grew, he said, they would inevitably be approached
by other industries, 'and at that point we have to stay true to who we
are and our moral compass.'

The conversation would stay with me beyond the story I wrote.
The more I thought about it, the more I decided Holmstrom's trouble
was that he needed a dose of the humanities and social sciences. He

was not a bad-faith actor, nor was he an idiot. It just seemed like this was the first time anyone had actually asked him how his ethics applied in the real world.

When he was forced to confront the issue, he approached it with the logic of an engineer looking at a machine. As far as he was concerned, the stuff with Santos was a small part of a greater whole. Proportionately, it was irrelevant, and certainly outweighed by the other good work the company was doing in bringing technology into classrooms across the developing world. If systems thinking was a way of looking at how a system's parts interrelate, this conclusion seemed reasonable.

Holmstrom clearly had genuine passion for his work; he wanted kids to have fun playing with robots and programming as they learned interesting things about the way the world worked – and he was making it happen. But whether he was willing to admit it or not, STEM Punks' actions had helped Santos get its corporate brand into schools. Holmstrom himself had also appeared on stage at a petroleum industry conference where he'd offered his thoughts on the most effective way to shape the attitudes of children to industry. His organisation even ran a workshop for high school students at the same conference. He hadn't done these things with some kind of ulterior motive – quite the opposite – but it was clear STEM Punks had served as a vehicle to deliver the Santos brand into schools. Sheer enthusiasm – and a desire to be part of the solution – was not enough to overcome this reality.

It was only after speaking to Holmstrom that I found Elaine Grossman's article outlining how the industry could better target children in schools. Reading over it, I was reminded of my conversation with the STEM Punks founder. I wanted to show it to him and ask whether it changed his thinking in any material way. Here was evidence that appeared to show how STEM Punks – an organisation whose mission he believed in – had been exploited. But when I tried following up, he never responded.

I did, however, show Grossman's article to Correna Haythorpe, federal president of the Australian Education Union. Her response was anger. The union's broad position was that there should be no corporate influence in schools. To that end, they had successfully fought a campaign to kill the Commonwealth Bank's Dollarmites program. What had been advertised as an altruistic endeavour to teach schoolchildren financial acumen turned out to be a cynical program designed to secure the next generation of customers. At one point, tellers had even fraudulently exploited the program by creating thousands of children's saving accounts to meet aggressive performance targets and earn bonuses.

What caught Haythorpe's attention was just how explicit the AIP–APPEA program had been in its aims. It was clear, she said, that what was being outlined was 'not about helping students, but about making sure the industry survives'.

'They talk about the importance of reaching young people with a clear message. That's brainwashing to me,' she said. 'They're saying we've got to get in and influence these young kids and make sure we can get them to understand that the fossil fuel industry is good, to make sure they work in it and to make sure they invest in it, into the future.

'It's an orchestrated campaign, right? There's an organising strategy here.'

HIGH PERIOD

9

Action Is Not Essential

In December 1996, Barry Jones, the new APPEA CEO, sauntered through the door and picked up the gun Dick Wills had built. He would go on to fire it point-blank into the heart of the Kyoto Protocol.

Jones – who shared a name with the Hawke-era science minister but was no relation – was, like his predecessor, a former public servant with an impeccable pedigree that gave him some authority on climate change policy. As an economics graduate of the University of Queensland, he had successfully applied to the Department of Foreign Affairs and Trade while working as a high school teacher in Brisbane. He then spent some time with the Department of National Development, where he worked first in regional development and then in petroleum. In those days, he had an office next door to Roger Beale, future head of the Department of the Environment, and his boss was Michael Keating – who shared a surname with the future prime minister but was also no relation – a future secretary of the Department of the Prime Minister and Cabinet.

In the mid-1980s, Jones joined the corporate policy division at the Department of Resources and Energy, where his responsibilities included greenhouse policy. There, he served with distinction. Gareth Evans, who headed up the department as minister for resources and energy, would recall his departmental team – including Jones – was 'of the highest competence and integrity'.

Even at that time, Jones was a doubter. In 1989, his service was rewarded with an appointment to the government embassy in Paris as the Australian representative to the International Energy Agency, the Nuclear Energy Agency and the OECD Agriculture Committee. It was in this role that he claimed to have attended the 1991 international climate conference in Chantilly, Virginia, which opened the negotiating process for a climate treaty, to be signed in Rio the following year. These negotiations would pave the road to both the Kyoto and Paris agreements by setting the ground rules for how to get there.

US President George HW Bush deliberately chose to host the meeting at a palatial red-brick conference centre in the backwoods of Virginia, not far from a Civil War battlefield, in an effort to keep out the pressure groups. It failed. Environmental groups still turned up, and oil, gas and coal lobbyists flew in to stalk delegates on the conference floor like they were hunting wild game. At later meetings, lobbyists were banned from the floor. It wouldn't actually change much, but until that time there was nothing to stop them directly pressuring negotiators to drop demands for language that might impose any sort of cost on fossil fuel producers.

Among them was Julian Roy Spradley, known as 'JR', a lawyer from Washington whose ten-gallon hat signalled his loyalties. When a Bangladeshi delegation finally grew fed up and confronted him about the realities they would face as sea levels rose, he reportedly responded by telling them, 'The situation is not a disaster.'

'It is merely change,' he said. 'The area won't have disappeared; it will just be underwater. Where you now have cows, you will have fish.'

Another figure present was physicist Brian Flannery, who ran climate modelling for ExxonMobil and acted as the company's liaison with the Global Climate Coalition (GCC), an industry alliance of multinationals that included Exxon, Texaco Oil, Peabody Coal, Ford, General Motors and Australia's own BHP. But it was the cigar-chomping Don Pearlman who would prove to be the conference villain. Pearlman was an American lawyer, a partner in a Washington-based law firm, Patton, Boggs & Blow, and former undersecretary in the Department of the Interior during the Reagan administration. The firm would be responsible for creating the GCC, with Pearlman – the 'high priest of the carbon club' according to Jeremy Leggett – running point. His work at these climate conferences involved coordinating with the Saudi Arabian and Kuwaiti delegations to wind back the terms of any agreement – in one attempt during an IPCC meeting in 1990, Pearlman infamously sought to have the very words 'carbon dioxide' stripped from a summary report. As one climate campaigner later remembered, he was a 'first-class lobbyist ... on the wrong side'.

Climate scientist Bill Hare was working with the Australian Conservation Foundation at the time and attended Chantilly with the Australian delegation as an environmental delegate; his role was to run interference against the fossil fuel lobbyists.

'It was my job to counter the bullshit,' Hare says. 'It was basically guerrilla warfare at a diplomatic level.'

Hare's work involved tracking what his counterparts working for the fossil fuel industry were doing, exposing these activities and countering them wherever possible. In an era before mobile phones and the internet, he and his team went so far as to raid the wastepaper bins at the end of each night looking for intel in the notes passed between delegations.

'What I didn't know when I started is just how bloody sinister they were,' he says. 'They were everywhere, and they were in your face. They didn't like what you were doing, and they weren't afraid to come up and shirtfront you on that. And you just had to stand your ground.'

It would have been impossible for a figure like Barry Jones not to come into contact with the fossil fuel lobbyists, Hare says. As a member of the Australian delegation, he remembers 'running into' Jones.

'He was a bit of a prick, really,' Hare says. 'He was very much on the fossil fuel side. He thought that we all needed fossil fuels, and this was essential to development and so on. He was very unhappy about the potential for climate action.'

After Chantilly, Jones was once more on hand as a member of the Australian delegation sent to the 'Earth Summit' in Rio de Janeiro. The event did not begin smoothly. President Bush had initially threatened to boycott the whole affair if the planned UN Framework Convention on Climate Change (UNFCCC) included a binding agreement to enforce emissions-reductions targets. When the targets were eventually dropped, Bush flew in to sign.

That was the first time the global oil and gas industry was able to marshal enough influence at an international climate conference to bend outcomes in its favour. Following the meeting, Australia became one of the first countries to ratify the UNFCCC, but cabinet documents from the period reveal that the Keating government struggled with the Australian paradox: the country was 'a major user and exporter of greenhouse-gas-producing fossil fuels and energy-intensive products' and had built a lucrative export industry selling coal to Asia since colonisation. Yet, at the very same time, there was recognition that Australia 'could be significantly affected by global environmental climate change'. On one side of that simple equation were the coalminers and the oilmen; on the other were the Australian people. Cabinet resolved the situation by amending a press release from the environment minister to emphasise the fact that the convention did not 'bind any signatory to meet any greenhouse gas target by a specified date'.

Australia's 'no regrets' policy on climate – the notion that Australia would only act on climate change insofar as this action did not hurt its economy – was already firmly entrenched at this moment of historic compromise. From then on, it became an article of faith within the public service and a basic tenet of Jones' worldview.

Jones was a man who believed in petroleum. According to one former colleague, he had a 'long history' working on petrol prices and a 'history in the department of working with the oil and gas industry'. His general philosophy was the same as the petroleum producers': taxes and regulation weren't the answer to climate change. He thought it better that the world innovate the problem away through technological advancement. It was an idea that would later be boiled down to a three-word slogan – 'technology not taxes' – under Prime Minister Scott Morrison. Renewable energy and electric cars were decades away, Jones reasoned. This meant the world needed oil and gas producers. In fact, the oil industry might even contribute to solving climate change by helping to develop carbon-capture and storage technologies – another industry talking point that would echo through the decades.

The only way forward, he declared, was to keep pumping the gas.

In 1999, simmering tensions within the industry saw gas producers walk out of the Australian Gas Association and join APPEA. The switch buttressed APPEA's influence just as gas began to play a more important role in Australia's export sector. Oil had long since given way to gas as the fossil fuel of choice among Australian explorers, but the boom was yet to land. The inclusion of the gas explorers did, however, rebalance APPEA's internal politics and approach to issues like climate change. As had been pointed out back in 1973, on the floor of the APEA conference, any direct threat to the ongoing business of oil and coal companies represented opportunity for their counterparts in gas – the 'cleaner' fossil fuel.

Jones might have preferred governments do nothing about climate change – but if they insisted on action, there was an opportunity to cast gas producers as heroes in the public eye. Burning gas for power generated CO_2, and the entire process, from the point of extraction to retail, was prone to leaks that amplified the effect of other greenhouse gases. None of this was acknowledged in industry talking points. Rather, the goal was to ensure people listening heard how gas burned cleaner than coal.

Its moniker – 'natural gas' – also helped. The term had been developed back when companies were switching out the dirty 'town gas' in people's homes, and it had stuck, even though natural gas was mostly methane, the second-worst greenhouse gas after carbon dioxide. Extracting this natural gas from the ground and refining it was easier than turning coal into 'town gas'; it was far more abundant than oil across the Australian landscape; and it could be found in conventional deposits – pockets of pure gas trapped beneath the surface – and 'unconventional resources'. Unconventional resources were sources from which it was more difficult to extract the gas, such as coal seams or clay-rich layers of rock known as 'shales'. Working these sources generally required pumping a chemical cocktail into the rock to fracture it and release the gas. When this is taken into account, gas was really no cleaner than any other fossil fuel. What mattered, however, was the name. As an exercise in branding, it clicked with the cynical logic of industry marketing teams. To be 'natural' was to be good. Forests were natural, as were rainbows and cuddly animals – and so was 'natural' gas.

Jones' innovation was to take things a step further and frame gas as a 'bridging fuel' between the coal-fired past and a sunny, renewable future. Wind and solar were in development, but they wouldn't be up and running for decades – certainly not without the levels of subsidisation his own industry had long enjoyed. In the meantime, Jones argued, gas could help clean up the grid, and the country could seek more oil to satisfy consumption in the interim. He even had a corny acronym to sum up this vision: UMGAFMO – Use More Gas And Find More Oil.

His was a view that would come to be shared by a coalition of other industry organisations, right as climate change became a flashpoint. The Australian Industry Greenhouse Network (AIGN) had begun life in the early 1990s, before Jones started at APPEA, as a loose industry alliance working on greenhouse policy. According to industry lore, the Mabo decision of 1992, in which the High Court of Australia overturned terra nullius, establishing the native title system of land rights, was the impetus for coalminers and related industry associations to get involved in AIGN. The broader resources sector considered the judgment a significant loss, as they hadn't been able to finagle a compact with government to stave off intervention. In effect, they thought it a total failure of their influence.

AIGN emerged as climate change became an issue for governments and industry groups began to cooperate. At first it served mainly as a policy forum, where its members could collectively work through government proposals about how to respond to the looming threat of climate change. Its membership was limited to a handful of core organisations representing energy-intensive industries, and its representatives would meet regularly to talk through the implications of greenhouse policy in a collaborative way. Early on, it was agreed that the network would not challenge the underlying science of climate change directly, as individual companies and their associations had in the past. Instead, it would focus its public communications on questions about how best to respond to the problem.

In 1994, AIGN's focus began to change, as it evolved from corporate study group into a vehicle for coordinating lobbying activity, with its own bank account, funds and AIGN staff donated by the membership.

Around this time, its member associations had early warning that the Keating government, led by environment minister John Faulkner, was considering introducing a carbon tax. This intelligence set off a flurry of activity within AIGN as the organisation began to coordinate a response. To broaden its representation, an effort was made to recruit other industry associations, and the network's membership expanded to a total of 14 organisations, representing Australia's heaviest polluters – notably coal, oil and gas producers, car makers, and the aluminium and cement industries. In short order, it became a steering committee for an alliance that spanned the entirety of Australian heavy and extractive industry.

When the interests of its membership aligned, AIGN was capable of pulling together the financial, technical and political firepower of its member companies and directing it all towards a singular goal. In 1995, that goal would be killing the Keating government's proposal for the first carbon tax.

AIGN set to work drafting communications plans, talking points, speeches and press releases, and commissioning research. Though APPEA and AIP were influential members, the Business Council of Australia was chosen to be the tip of the spear and lead meetings with government.

By far the most consequential decision of the campaign was to loop in the Australian Council of Trade Unions (ACTU). However it came about, the ACTU's climate change spokesperson, Tony Wilks, would later justify the decision to join the campaign with a string of arguments that would be familiar today. A carbon tax would hit those on the lowest incomes the hardest. Placing a burden on the country's coal and gas exports was a bad idea, he added, particularly at a time when no other country was implementing the same measures. And then there was the risk of job losses:

> If industries such as the coal industry, aluminium industry and so on became less competitive than they are, and they start to dry up, there's nothing to replace them with. Many of those employees in those industries are in remote areas. There's a huge structural adjustment cost associated with moving people out of those areas [and] retraining them. What do you retrain them for? Where do you send them to work? So you start to build up a picture which is just too scary to contemplate from a public policy position.

With the combined might of business, the unions and every heavy industry in the country lined up against the proposal, a fourteen-foot 'wall of opposition' was erected, with hundreds of millions of dollars in financial resources at its disposal. Faulkner didn't stand a chance.

Confronted with this hostility, the government cast around for an alternative – and Barry Jones was happy to supply one. Before APPEA, Jones had headed up the Pulp and Paper Manufacturers Federation of Australia (PPMFA). Forestry was considered a training ground for CEOs in the extractive industries, with APPEA the next rung on the ladder. During Jones' tenure, the organisation made submissions to the senate Environment, Communications, Information Technology and the Arts references committee outlining how it thought things should go:

> The PPMFA considers that an expanded program of voluntary agreements should form the foundation of Australia's national

greenhouse response. In our view there are many potential 'no regrets' and 'low regrets' measures that have yet to be fully exploited. This should be done before other measures of a more mandatory nature are considered.

It was a suggestion that would be taken up. Rather than mandating change through legislation, the government of the day settled on a voluntary 'early action' initiative called the 'Greenhouse Challenge'. Individual companies or their industry associations would be encouraged to sign up. Once in, they would reduce their emissions by an agreed amount, a commitment that would then be reported to the Australian Greenhouse Office. APPEA would be among the first associations to sign on to this accord as a good-faith gesture. Whether the association acknowledged it or not, the program conveniently helped blunt more stringent government intervention. The agreement struck with government covered 98 per cent of the Australian oil industry, including well-known entities like BHP Petroleum, Esso Australia, Mobil Exploration and Production Australia, Shell Australia, Santos, and Woodside Offshore Petroleum.

Later, APPEA would claim this as proof the association and its membership were committed early on to climate action. In 2000, however, it would be found only eight out of 76 participants in the entire program had met their original forecasts. In 2002, APPEA's members claimed to have made a reduction of 21 million tonnes of CO_2-equivalent, but it was impossible to assess the veracity of this claim. Critics pointed out that the agreements were secret, and many of the reductions were efficiency gains that would have been made in the ordinary course of business anyway.

What made for great marketing turned out to do little to contain Australia's emissions, which continued to rise. Having spun themselves as heroes, Australia's oil and gas companies had chalked up a win, although it was only a warm-up for what was to come next.

★

From 1946, the year Barry Jones was born, to when he took over APPEA in 1996, the amount of CO_2 Australia pumped into the atmosphere each year had grown tenfold. Over the course of his eight-year service with APPEA, Australia would add another 74.26 million tonnes of CO_2 to its yearly total, twice the amount released in the year of his birth.

Jones would oversee this increase with pride, considering it a mark of progress. If Dick Wells had been quiet and measured, Jones – according to those who worked with him – was 'bombastic', fond of reminding staff that APPEA under his leadership was 'not a participatory democracy'. In government, he had never risen higher than band one in the Senior Executive Service. Having been overlooked for promotion, once out in the private sector he relished power and gave no quarter in his interactions with the bureaucracy. A former colleague recalled how, upon returning to the office, he would be overheard to say he 'loved bashing public servants', without any apparent acknowledgement that he had been one himself.

With Jones at the head of APPEA, Dick Wells in charge of the Minerals Council of Australia and Keith Orchison heading up the Electricity Supply Association of Australia, the three formed a core troika within the broader industry alliance.

By this time John Howard had taken government; behind the scenes, AIGN continued to lobby aggressively on the 'greenhouse issue', often well out of view of the public. Thanks in part to Jones' influence, AIGN's objective at this time was to prevent any hard constraint on greenhouse gas emissions. Failing that, it sought to delay the introduction of any constraints as long as possible. If it was unsuccessful in that, the fallback position was to demand exemptions or compensation.

Jones is credited with putting AIGN on a war footing. Though its membership had taken a constructive approach to the issue early on, its work during Jones' tenure was redirected towards drawing up orders of battle. A brief he published in the July 1997 edition of APPEA's members-only newsletter reveals his thinking about climate change ahead of the international negotiations that created the Kyoto Protocol. In it, Jones outlined AIGN's position in four key bullet points:

- The Berlin Mandate and the Framework Convention on Climate Change clearly recognise the need for equitable outcomes;
- Measures that go beyond 'no regrets' policy are not justified by the current state of the science of climate change. The science does not make it clear that there are a number of time paths to arrive at the same outcome, so action is not essential – it's a time-of-action versus cost-of-action trade-off that has to be made;
- Climate change is a global problem requiring a global solution. Action only by developed countries will have little or no environmental effect;
- There are real doubts about the capacity of the European Union, the USA and Japan to deliver and about their intention to deliver.

What this document establishes – in Barry Jones' own words – is that APPEA, the core of the Australian oil industry, did not just seek to cast doubt on the science of climate change in 1989, but, through its involvement in AIGN, it was part of an active campaign against government efforts to address the problem.

In disclosing AIGN's internal thinking on climate change to the APPEA membership, Jones confirmed the existence of a reactionary political program through which Australian industry sought to stop climate action, or at least buy itself time. In late 1995, the IPCC had published its second assessment report, which said that the statistical evidence pointed to 'a discernible human influence on global climate'. As the authors of the report had faced massive resistance from the Global Climate Coalition, Jones' brief suggests that AIGN was mounting a rearguard action to keep opposition alive locally in Australia – though it remains unclear what, if any, relationship AIGN had with the GCC.

But Jones was not done: in the same brief he also listed a series of 'concerns' held by AIGN and its member organisations – a rare, candid explanation of the opaque organisation's goals and motives. In this list was an explicit statement of AIGN's plan to delay action to address climate change as long as possible.

'The longer the adjustment time frame for meeting any targets the better. This will allow for the capital stock to be run down and new technologies to be introduced,' Jones said.

By 'capital stock', Jones meant every dirty, inefficient factory, smelter, refinery, shipping terminal, drilling rig, power plant and platform in the country. What he was calling for was a delay in any transition, to give Australian industry a chance to run their operations into the ground and extract every last dollar they could before cutting their workforce loose.

It was a scientifically illiterate position. Every molecule of carbon dioxide or its equivalent stopped from entering the atmosphere reduced the potentially catastrophic effects of climate change. Delaying a phase-out in order to allow fossil fuel producers to get their financial house in order would result in measurable change within the atmosphere.

But Jones went further still. He also suggested it was AIGN's view that – assuming harm had already been done – a cost-benefit analysis showed there was no point in acting on the problem.

'Adapting to climate change may be a cheaper option than emission reduction,' Jones said. 'Adaptation has to be considered since, if the science is correct, some change is inevitable.'

Not only did these statements demonstrate a clear knowledge or awareness that, on the basis of the science, harm was probable, they were also an explicit sign that AIGN and its member organisations would seek to continue their potentially harmful activities because 'the damage had already been done'.

Jones was equally clear in relaying AIGN's view that 'Australia should not sign on' to the Kyoto Protocol.

'All policy options must be kept open and none made mandatory,' Jones said. 'A legally binding target is unacceptable. No-one knows what a legally binding collective target means.'

This position was not necessarily shared by all of AIGN's members, and some groups would eventually drop out, but those counted among its ranks in 1994 included the Australian Aluminium Council, the Australian Automobile Association, the Australian Chamber of Commerce and Industry, the Australian Coal Association, the Australian Gas Association, the Australian Institute of Petroleum, the Australian Mining Industry Council, the Australian Petroleum Production and Exploration Association, the Business Council of Australia, the Cement Industry Federation, the Electricity Supply Association of Australia, the National Farmers' Federation, the Plastics and Chemicals Industries Association and the Pulp and Paper Manufacturers Federation of Australia.

In other words: the entirety of Australian industry.

On climate change at least, the overlap between government and industry in Australia during the mid-1990s can only be described as a grim period of state capture. The industry's political manoeuvring was not so crass as in developing countries. Australia is a wealthy nation in which process matters and crude displays of power are not tolerated. The best way to a desired outcome is to control the process itself, or, at the very least, specific choke points in that decision-making process. At the peak of its influence, that's what AIGN was able to achieve.

Right as AIGN was organising to pressure the government into rejecting the Kyoto Protocol in 1997, a small group of business leaders was busy working to make sure the interests of coal and gas companies were hardwired into any future decisions about the direction of the Australian economy. The Howard government had commissioned a review of the future of Australia's economy and stacked its advisory body with friends and allies. James Hardie director Meredith Hellicar, a former executive director of the New South Wales Coal Association, sat on the committee, as did multimillionaire Harold Clough, a Liberal Party financier, Institute of Public Affairs board member and climate denier whose engineering firm had made a fortune selling picks and shovels to oil and gas producers. The white paper they published carried the title 'In the National Interest' and sought to entrench the position of the nation's fossil fuel producers.

'Australia is a leading exporter of energy, especially coal and liquefied natural gas, reflecting its comparative advantage and reliability as a supplier,' it said. 'The Government will work to ensure that Australia maintains these advantages.'

In many ways, this statement only formalised a view which had already taken shape within the Department of Industry and the Department of Foreign Affairs and Trade a decade before. What it shows, however, is the extent to which industry – fossil fuel producers in particular – were able to influence processes to shape certain outcomes.

Among mates, those within AIGN bragged about what they were doing. When Guy Pearse, a former Liberal staffer, interviewed all the key players for his PhD, he found they referred to themselves as the 'Greenhouse Mafia'. Investigative journalist Marian Wilkinson would, borrowing a term from writer Jeremy Leggett, dub them the 'Carbon Club'. What defined them was their level of access and close coordination. A core feature of the organisation was regular, monthly members meetings with politicians, bureaucrats or thought leaders. In addition to these official engagements, Pearse learned, the early AIGN leadership met for 'executive directors meetings' every four months. Keith Orchison convened these meetings, which were attended by Barry Jones, Dick Wells and others. Climate change – referred to as 'greenhouse' – was always on the agenda as they discussed their shared issues and 'looked for synergies'.

Protected by anonymity, Pearse's subjects were not shy about explaining their intentions. Their actions were directed towards 'fixing the outcomes' to industry's benefit. Ordinarily, lobbying in Australia involved working your way up the chain of command until you found someone with authority who might be spurred to action. If the policy adviser wasn't listening, you'd try the departmental secretary. If the secretary wouldn't move, you'd try the minister's staff – or better yet, the minister. When a state government waved you off, you'd try the feds.

This dynamic was captured by Pearse in one celebrated war story offered up by an industry insider, featuring David Buckingham, CEO of the Business Council of Australia; Dick Wells, then CEO of the Minerals Council of Australia; and Arthur Sinodinos, John Howard's chief of staff:

> We used to spend a bit of time to making sure that a whole bunch of CEOs that were members of the [BCA] used to go to all the policy meetings of the BCA because David was quite a clever operator. David's style would be to talk in generic terms about – to use terms like, 'I had it from the highest levels of government that ...', and people would say, 'Who, David?' And it would be some adviser or whatever. And David would interpret, you see. So, I can remember a celebrated meeting of the Minerals Council where he tried to do this, and there was Keith Orchison and Dick Wells and various others there ... And Dick just excused himself. David had said 'I had from the highest levels of government that government is concerned that industry expresses a view about greenhouse etc etc.' And he was using that as a driver to drive Australian business in a certain organisational direction. So, Dick picked up the phone and spoke to Arthur Sinodinos, because he presumed that [David was] talking about him since he said 'the highest levels of government'.
>
> [...] So Dick calls Arthur, and he said, 'Arthur, Buckingham is sitting in a room next to me in my office here telling us that government wants us to do this, this, and this. And he is talking like it is coming from you.' And Arthur says, 'Well, it has not come from me, and we do not want you to do it.' And so Dick walked back in and said, 'Look, sorry, David – I just talked to Arthur Sinodinos and he disagrees completely with what you just said.' It was that sort of game. You see, David's stupid presumption was that he was the only person that could access high levels of government. Of course we all do.

AIGN's reach, however, lay not just in its direct access to power but its influence on the machinery of government itself, thanks to the shared financial and political resources of the industry alliance's combined might. Having hired from within the ranks of the private sector and government, groups like APPEA had purchased the corporate memory of these institutions wholesale. Their employees knew more about the intricacies of climate policy than those working for their former departmental employers and were sometimes called upon by their successors to help draw up documents. One of Pearse's interviewees, an industry lobbyist, described both having 'sat inside [a federal department] and helped to draft cabinet submissions' and having vetted documents before they were sent to cabinet for discussion.

It was a far cry from the early days of the Australian oil industry and its ham-fisted stab at influence. Though it may have started from behind, industry's penetration of government processes in Australia was much more advanced than that of its counterparts in the US. One interviewee who spoke to Pearse boasted: 'In the US, they sit in the gallery – in Australia, they sit in the room. They are part of the team.'

How these people used their enormous influence to shape greenhouse policy, stall for time and shut down action from the late 1990s through to the mid-2010s, is a story that has been told many times, in forensic detail, by others like Pearse, Wilkinson and Clive Hamilton. At this point, it is the stuff of Australian political folklore. In short, John Howard would sign the Kyoto Protocol with his fingers crossed behind his back. This agreement proposed the first binding international emission reductions targets and groups like AIGN threw everything they could at stopping it being signed. The Department of Foreign Affairs and Trade essentially viewed climate talks as another trade negotiation; in the background, a coalition of climate deniers, front groups and Liberal Party financiers railed against the 'economy-killing' treaty in the press. Exxon even partly funded its own astroturf campaign, holding a conference in August 1997 titled 'Countdown to Kyoto'.

Despite these efforts, Australia signed the document the next year. Thanks to cynical diplomatic manoeuvring, the 'Australia clause' was written into the agreement first, in an eleventh-hour coup – creating an accounting loophole that would allow Australia to claim credits for not clearing as much land as it otherwise would have. Upon his return, Howard's environment minister Robert Hill received a standing ovation in cabinet.

Before long, it became clear Howard had no intention of ratifying the Kyoto Protocol – incorporating the international treaty into

Australian law to give it practical effect – unless the US did so first. This was foreshadowed in a private briefing given to AIGN members by minister for resources and energy Warwick Parer in 1998. Parer, an avid defender of the coal industry, a climate change denier and a former housemate of Howard, was later discovered to have been holding $2 million in coalmining shares. The details of this meeting were splashed across the front page of the *The Canberra Times*. Apparently, Parer had told the gathering of industry figures of a secret cabinet decision not to ratify the agreement unless the US government did so first. Leaked meeting minutes recorded APPEA's Barry Jones saying to Parer, 'We can have that in writing? [...] That is a resoundingly positive statement.'

Howard waited until 2002 to make the non-ratification official. As Wilkinson reported, Howard appeared to have made his decision the year before but was waylaid in announcing it by the September 11 terrorist attack, which took place while he was visiting New York. On World Environment Day, 5 June 2002, he stood on the floor of parliament and officially declared: 'It is not in Australia's interests to ratify the Kyoto Protocol.'

The decision would isolate Australia internationally for the next five years, but it bought industry time. The Howard government claimed it would go it alone on greenhouse – that Australia could find its own way to lower emissions – but none of the ideas it tossed around internally found traction. Either they had no effect and were a waste of money, or they had an effect and were immediately opposed by those looking out for the interests of fossil fuel producers. The result: nothing got done.

The years 1996 to 2003 was the high watermark of AIGN. APPEA would remain strong for the next two decades, but it would never achieve the same level of finesse on its own.

Still, it had all worked superbly. It would be another four years before the Kyoto Protocol was ratified. Barry Jones, however, would never get to live in the world he helped build. He retired suddenly in 2005 and died only a few months later.

★

When Kevin Rudd was elected prime minister in 2007, it was a profound shift in the political landscape that brought with it a change in tone at AIGN. On the new Labor government's first day in office, Australia ratified the Kyoto Protocol. The country now had two targets: to limit emissions to 108 per cent of 1990 levels by 2012, and to cut emissions by just 0.5 per cent of 1990 levels by 2020. It was not enough, but it represented a shift in approach.

The shine had worn off AIGN by then – it no longer had its hooks so deeply in government – but it persisted. It was there throughout the prime ministership of Julia Gillard, whose attempt to introduce a carbon pricing scheme made her a favourite target for misogynist abuse by climate deniers. It was still there during the tumult of the Tony Abbott years, and survived his replacement by Malcolm Turnbull, who, took the risks posed by climate change seriously and believed the matter demanded action.

Its lobbying power may have waned, but AIGN continued to claim special access, offering its members monthly, off-the-record meetings with decision-makers. These meetings followed 'Chatham House' rules: the participants' identities were confidential, but information disclosed could be freely used. The organisation's 2015–2016 annual report described these events as a key benefit of membership:

> The monthly network meetings are the focal point for these exchanges, providing members a Chatham House forum to test theories and engage with government and opinion leaders on the adoption of a principled national and international policy framework within which effective and equitable domestic climate change policy is developed and implemented.

In 2018, Malcolm Turnbull would in turn be ousted by his own party, denounced as a socialist for supporting climate action, and replaced by Scott Morrison. Morrison's rise secured something of a renaissance in their fortunes. He was the man who brandished a lump of coal in parliament – supplied by the Minerals Council of Australia – and committed the country to a "gas-fired recovery" following the global Covid-19 pandemic.

Throughout all those years, the industry would continue to foster close links with elected officials and their staff on both sides of politics. Barry Jones had been succeeded by Belinda Robinson, who previously worked on environmental policy within Department of Prime Minister and Cabinet. As of 2021, APPEA has counted several former staffers to state and federal Labor and Liberal ministers on its payroll at one time or another. Among its alumni are an appointee to the Productivity Commission and a former federal minister who served for a time as its chair. Damian Dwyer, APPEA's former executive director, would join AIGN's board in 2012 and go on to serve as AIGN chair.

AIGN's membership still includes 17 companies and seven industry associations, but its influence has ebbed. As of 2024 the organisation has officially embraced the goals of the Paris Climate Agreement – but its historical advocacy created a lost decade in Australian politics, colloquially dubbed the 'Climate Wars'.

'They won, basically,' says energy transitions scholar Marc Hudson. 'They were always going to win, but they did it with aplomb. We have the wind and we have the sun, but we also have the coal and the gas, and they wanted to keep making money.'

10

Hot Air and the Cold Facts

It was one of those mysteries I couldn't resist: a small anecdote buried deep in the footnotes of a sprawling document filled with a hundred more immediate outrages. Depending how you framed it, the story either amounted to little more than interesting historical detail or a shocking tale of corruption in which information about climate change was concealed from the public to protect fossil fuel interests.

The detail was contained in Guy Pearse's PhD thesis. It had been 20 years since it was published, but it was still full of leads yet to be followed up. Having gathered together his research, Pearse had done something unconventional in an academic work. So shocked was he by what he found – and so unbelievable were the things his interviewees discussed so openly – he had chosen to publish whole slabs of raw transcript from his anonymised interviews in the footnotes.

These vignettes offer an insight into those responsible for undermining the country's efforts to meaningfully address climate change. It wasn't possible to know who had said what, of course; Pearse had promised to keep his sources' identities a secret, as is required by university ethics committees, but what mattered was that their war stories were recorded. Pearse's research offered a

tantalising peek at those who had mortgaged our collective future
for a quick buck.

The story that caught my eye appeared in footnote 645. Whoever
was speaking had been trying to disabuse Pearse of the notion that
government had been captured by fossil fuel producers, saying it
was a 'myth' that the business community 'had a hand up [the
government's] back' and treated it 'as a glove puppet'. The setting
was a gathering of '200 or 250 captains of industry' convened by
minister for resources and energy Gareth Evans in 1986. The aim
of this meeting was to plan the direction of Australia's long-term
energy policy all the way to the year 2000. The result, Pearse's
informant explained, was a policy document titled *Energy 2000*:

> By the time the product of this thing was published Gareth had
> moved on, and if you go and get a hold of a copy of *Energy 2000*,
> which you can from the Parliamentary Library, you will find, and
> I forget how many chapters it is – I think it's 12 chapters – it's very
> obvious that one's missing, and the one that's missing is the one on
> greenhouse and the reason the chapter on greenhouse is missing is
> that the then senior public servants perceived it as their patriotic
> duty to prevent the coal industry from being undermined by an
> untoward focus on something that in their thinking was a load of
> cobblers.

What this interview suggested, is that not only had public servants
engaged in an act of censorship, but the informant also claimed they
had acted alone, telling Pearse: 'I don't think you could even argue
that it was because they were under intense lobbying pressure from
the coal industry.' Instead, they added:

> I think it was very much a matter of some senior and quite strong
> public servants taking it into their heads that having a whole
> chapter in something like this on greenhouse was just plain wrong,
> so they took it out, or they persuaded the minister of the day who
> was Peter Cook.

At the end of the interview, the informant confirmed that the chapter had been written, though they had 'never seen the aforesaid piece of paper'.

'It was in the original draft and it got removed,' they said.

If true, these were shocking allegations. Pearse's source was claiming that senior Australian public servants had been so loyal to fossil fuel interests that they had acted on their own, without any pressure from industry, physically removing direct references to climate change from a public document because it might compromise the business interests of coalminers. Whoever was speaking might have considered this proof that industry didn't control government, but I interpreted the events they described very differently. It seemed to me that oil, gas and coal companies had been so good at controlling how, when and what information was available to the public service that decision-makers could not even imagine an alternative point of view. These public servants were so conditioned to this line of thinking that they had weighed the potential outcomes, divined the industry's needs and then moved to neutralise any threat long before it became a problem – all, allegedly, without being asked. It was an extraordinary tale, one which raised the question: if this had actually happened, how many other times had similar acts occurred?

I immediately began to investigate. It took all of five minutes to confirm that Gareth Evans had convened a meeting of 250 industrialists in 1986. When I cross-referenced discussions about what occurred at that meeting with records kept by the oil industry itself, they described a hostile encounter. Both APEA and AIP were there, as were the coal producers. The conference had been the impetus for these organisations to cobble together an ad hoc alliance of industry groups and corporate executives to oppose any proposal that might increase the tax burden on their collective membership. The more I read, the more this informal alliance sounded like a precursor to what would later become AIGN.

Having established all this, the next step was to find out what the *Energy 2000* policy document held in the Parliamentary Library actually said. Problem was, as I was not a staffer, I had no access to it. What I do have, however, is friends in both high and low places. With their help, I found a workaround and was able to view a copy of a document that had been circulated to industry in 1986 for comment and feedback.

Taking a look at the actual document suggested inconsistencies in the account given by Pearse's informant. This version had ten volumes – not twelve chapters. A volume on environmental issues made no mention of the greenhouse effect, but there was a reference in the part dealing with coal:

> The so-called 'greenhouse effect', thought to be caused by increased carbon dioxide levels resulting from fossil-fuel combustion, will all require consideration by governments and industry. These problems have little effect on the domestic use of coal, but they could affect the use of and cost of using coal overseas, with consequent implications for our exports.

Not only did this frame the greenhouse effect as subject to doubt, but it sought to understand the problem only in terms of what it might mean for Australia's fossil fuel exports.

'Australia is fortunate in having large quantities of high-quality, low-sulphur coals, and this could be a marketing advantage in the more densely populated areas of Europe and Asia,' it said.

It was a single mention in ten volumes of text, but finding it made me wonder if Pearse's informant had been wrong. The documents – the 'Red Books' as industry and government sources called them, owing to their red covers – may not have discussed climate change with any real insight, but they *did* talk about it. A closer inspection gave no indication there was a missing volume; Pearse's informant said it was 'very obvious' a chapter was missing. It appeared they were mistaken.

For a time, I put aside this research. I forget about it until much later. Then, hunched over a desk at the National Library in Canberra poring over old industry materials, I found a box of records. When the 1986 version of the *Energy 2000* policy documents were first circulated, there was a period of consultation during which those concerned could weigh in. This box contained copies of the responses

the companies had sent back. Among them were tightly bound
tomes from Shell, Esso and the Australian Institute of Petroleum
complaining about proposed taxation changes. All standard stuff. But
the box also contained a letter from a CSIRO scientist named Roger
M Gifford, typed across two thin A4 pages. Stapled to these delicate
pages was a photocopy of a public statement, which he had helped
produce, from the Villach climate conference in 1985 – one of the
first international climate conferences – stating that temperatures
in the 21st century would be 'greater than any in man's history'.
Gifford's letter, on CSIRO Division of Plant Industry letterhead,
was addressed to the secretary of the Department of Resources and
Energy, and it upbraded the department for its failure to address 'a
topic not covered that is too important to ignore in the timeframe
to 2000 AD'.

A key sentence had been emphasised with neon-orange
highlighter: 'This is the impact of the globally increasing atmospheric
CO_2 concentration owing to fossil fuel burning.'.

He went on to explain how, as a major coal exporter, Australia had
a responsibility to act on climate change, given the CO_2 emissions it
had helped create over 150 years of coal exports. For good measure,
Gifford concluded his letter with a curt warning: 'Atmospheric
carbon dioxide cannot be ignored in energy policy from now on.'

Later, after I flew home, I went back to search again, this time
looking for the second, final edition released in 1988. Conveniently,
a copy of it had been digitised and was available online, complete
with searchable text.

Pulling up the document on my laptop, I scrolled through it. A
foreword signed by Peter Cook, who served as minister for resources
and energy in 1988, described *Energy 2000* as a 'comprehensive guide
to the current state and future uses of our energy resources'. It ran
15 chapters long, with a sixteenth chapter chock-full of appendices,
and not a single one dealing with the greenhouse effect. A keyword
search confirmed that the words 'greenhouse' and 'emissions' did
not appear once. Neither did the phrase 'climate change'. 'Carbon
dioxide' was mentioned once – but only in a section describing gas
as 'one of the cleanest fuels available'.

In other words: any reference to scientific concerns was gone. At some point between 1986 and 1988, when the final version was released, any mention of climate change, or the greenhouse effect as it was then known, had been stripped out, clean. If it had been handled in a dedicated chapter, those pages simply never made it to print.

I wanted to know more, so I contacted Gareth Evans to ask if he was aware of what happened to the policy papers after he left the ministry, but he said he could not recall. Colleagues of his, such as the science minister Barry Jones – 'a uniquely far-sighted intellectual' – did talk about climate change, Evans said, but 'his concerns did not have any resonance' in cabinet. For good measure, Evans contacted Tom Spurling, who served as his private secretary in 1985, to double-check his memory, and Spurling responded, saying the main policy issues at that time were 'still about conventional fuels'. Spurling added that there was considerable difference of opinion about climate change at that point and directed me to a statement made by the IPCC in 1990 that suggested a global increase in temperatures could still be a natural process. Both agreed climate change hadn't been on the radar, and that there was no 'hidden' chapter.

It was only after speaking to Roger Gifford that I learned what happened. He was pleasantly surprised to get my email, having written his letter nearly 40 years earlier. Back in 1985, he explained, there had been a fight to get governments to take climate change seriously. Now retired, Gifford struggled to recall all the details, but he was present at the initial conference – one of only a few voices from the scientific community in attendance – where he remembers questioning a senior person in the coal industry who had just given a presentation. When Gifford asked the coal industry figure what they planned to do about all the carbon dioxide their operations were adding to the atmosphere, he responded saying they would 'scrub the CO_2 out of the stack gases just as we have when we found we had to scrub the SO_2'.

'To me, that meant he had no idea of the scale of the amount of carbon dioxide in the atmosphere, which in mass exceeds the amount of coal,' Gifford said. 'That was very depressing for me, because this heavyweight, who was quite cocky, said that of course industry would handle it if society and the politicians said they had to.'

Having been given a chance to review an early draft, Gifford spoke with the late Dr Hugh Saddler, a friend and colleague who had been brought in as an energy consultant to work on the *Energy 2000* strategic plan. Talking it over, the pair thought similarly. Climate change had been overlooked, they agreed. It was an oversight that needed to be addressed.

'Hugh inserted several paragraphs on the topic, which I thought were good and well-balanced,' Gifford says. 'Then we sent that off for consideration by the departmental people drafting the documents.'

'It was with amazement when I saw that the [red book] documents had had all that removed,' he says. 'I was really pretty horrified and wrote a letter to the Secretary of the Department saying that it was a mistake'.

That would be the same letter I found in the National Library, though Gifford says he did not highlight the sentence in the text – that was likely done by an official trying to argue the case with the departmental secretary. What's more, he was not aware there had been *another* final version of *Energy 2000* printed which didn't mention climate change at all.

Without the missing chapter or the account of someone who worked on it directly, it was no smoking gun, but it was enough evidence to confirm Pearse's informant had been broadly telling the truth. Someone, it seemed, had deliberately suppressed or concealed information about climate change to protect the coal industry. Whatever had happened with *Energy 2000*, what I was looking at, I thought, was one of the first, perhaps accidental, wins by oil, gas and coal producers in Australia in their campaign to conceal the reality of climate change, a victory that would be replicated again and again in the following years – and all in the name of freedom.

★

At its heart, Big Oil's campaign to undermine the science of climate change was a fight for control of ideas. On one side of this information war were the oilmen with all the money, on the other were those earnestly trying to communicate a growing problem to the general public. The battleground would be the academic journal, the printing press, the reporter's notebook, the television camera, and later, the internet; the prize was the minds of the public.

This was a fight that didn't necessarily begin with climate change. The first shots were fired decades earlier, way back in the 1970s, a time when wealthy men were struggling to make their voice heard. In Australia, at least, they had just lived through the Whitlam era and were growing nervous about what could happen when the public stopped listening to them. What they wanted – what they felt they needed – was a way to get the average Australian to think like them. It would be an Englishman named Anthony Fisher who would show them the way.

On 19 August 1976, Santos co-founder John Langdon Bonython wrote a letter of introduction to a friend in Texas introducing Fisher. Bonython had come into John Murchison's orbit through his association with the Murchison family's patriarch, Big Clint Murchison.

Bonython, himself the privileged son of a family whose wealth originated in mining speculation, initially thought Clint Murchison a kindly old man, but would be warned by Reg Sprigg not to underestimate him. Texan oil millionaires had a reputation for being loudmouthed, opinionated and cheap – and Murchison ticked every box. One of the original Texan oil barons, Murchison – 'Murch' to his friends – had ridden to Bonython's rescue when his oil company was beginning to sink. Bonython had founded South Australia Northern Territory Oil Search (Santos) in 1954 at the suggestion of high school friend Robert Bristowe, but the company initially struggled to get off the ground. Cash-strapped, and lacking both technical expertise and equipment, Bonython began to look overseas for a partner. He would find one in Murch, thanks to a former student of a Santos geologist who had gone to work for the Delhi-Taylor Oil Corporation over in the US. Through this connection, and Reg Sprigg's legwork, the companies hammered out a joint venture, and in 1958 Murchison opened an office for his Australian subsidiary in Adelaide.

The Delhi-Taylor Oil Corporation was the only publicly traded family company that was overseen by Murch himself – but oil wasn't his only going concern. Oil might have brought the family wealth, but it was politics that secured it. A formidable figure in the right wing of the US Republican Party, Murch was a committed ideological warrior and prominent member of the exclusive, members-only Dallas Petroleum Club – the kind of place where a guy could grab a club sandwich and a billion-dollar deal. He boasted a close personal friendship with FBI director J Edgar Hoover and watched the McCarthy-era prosecutions with glee.

Santos didn't exactly thrive over the course of their partnership. Bonython wasn't particularly interested in the health of the business – at one point the Australian subsidiary of Murchison's company was caught billing Santos for expenses to cover a racehorse, Sweet Bippy. Neither is it clear where or how Bonython and John Murchison became friends, but when old Clint Murchison died, they stayed in touch.

Writing to John, Bonython began by informing his correspondent that it had been a 'sad time of recent years' in Australia and expressed relief at the fall of Whitlam and the end of his 'very regrettable Labour-Socialist [sic] Federal Government'. Coming to the point, Bonython informed Murchison that he was not writing to talk oil but to make an introduction. He had recently invited a man, Antony Fisher of London, on a two-week speaking tour to address company directors in Adelaide, Melbourne and Sydney. It was one of two tours Fisher would make that year. Bonython reported to his Texan counterpart that Fisher was the founder of an organisation called the Institute of Economic Affairs (IEA) and 'a successful businessman, youngish but retired', who was now being asked to set up similar institutes across the world.

At that time, Fisher – 'an intense, ascetic man who had been to Eton and Cambridge', according to one BBC write-up – was travelling the world with a copy of the Rich List in his back pocket, methodically arranging meetings with each person on it, ticking off the names as he went. Fisher himself came from wealth; his family made their money in mining, and he had made a second fortune from factory farming. He thought for a time about going into politics, but at a meeting at the London School of Economics in 1945 he was told by Friederich Hayek, an economist whose ideas laid the foundation for neoliberalism, that it was a waste of time.

'He explained that the decisive influence in the great battle of ideas and policy was wielded by "second-hand dealers in ideas",' Fisher later recalled.

'If I shared the view that better ideas were not getting a fair hearing, his counsel was that I should join with the others in forming a scholarly research organisation to supply intellectuals in universities, schools, journalism and broadcasting with authoritative studies of the economic theory of markets and its application to practical affairs.'

Taking this advice to heart, Fisher would arrive at Bonython's doorstep with a curious proposition: he had developed a method by which rich businessmen could inject their ideas into public discussion without revealing themselves as the source of those ideas, and he wanted Bonython's backing.

It was a proposition that captured Bonython's imagination:

I cannot tell you his method quickly. What *you* may say, what any *business* itself may say, is put down by many to 'vested interest'. He has a technique of getting academics to say and write under their own names what business cannot say for itself. It is now having some good results in the UK and in Canada. I have had, over the years, a good bit to do with Chambers of Commerce, of Manufacturers, with politicians – and what they say is regarded sceptically by a public encouraged by leftwing academics.

Fisher's method seems to me to be the best I have come across.

Sporadic attempts to defend private enterprise, private property, freedom of choice must, of course, be made whenever possible. However, Fisher's method is not so sporadic.

It is a continuing process. It provides ideas and books read by academics, students, journalists and the public. Also the books are reviewed in the press, and go some way to get on to the bookstalls and into the shops, something vastly different from the socialist nonsense that has proliferated for years. (The method can be backed up in many ways – by a society etc. etc.)

Fisher's 'method' boiled down to the creation of public relations shops for seeding ideas among the public. These 'think tanks' laundered the ideas of wealthy businesspeople like Bonython, promoting them through the work of other individuals with a veneer of credibility. Academics, journalists, policymakers – it didn't really matter whose work it was, as long as it supported the businessperson's point of view. Nor did it matter whether the 'expert' in question was in on the grift or a sucker. In fact, they didn't even need to be an expert in the specific area they were talking about. Journalists were obliged to present the arguments of authorities on 'both sides' of a debate, and think tanks exploited this demand for balance. If Fisher had anything to say about it, reporters wouldn't be left to seek out the facts on their own. He would make sure they got them.

The basic process was straightforward. A report, study or analysis was printed and shotgunned out into the world. Interviews were then solicited in print, on radio and television, particularly from any outfit that was syndicated, to ensure the key quotes travelled widely. Because these ideas would then appear to originate from different experts at different organisations, it created the impression of debate, when in fact it could all traced back to one source. It was a process that could be deployed in service of a particular idea, like free market economics, or to generate confusion about an issue, stoking doubts or creating contention where there had been none. Confronted with this confected uncertainty, decision-makers would find themselves suddenly unsure of their footing and nervous to act.

Exposing this after the fact had little effect on the outcome: by then, it was already too late. As Fisher understood early on, the first lie wins. No correction or pushback would generate the same headlines as the original story – and the resulting controversy often served to promote it further. Do this often enough, do it consistently over a period of years, or even decades, and the ideas you want to seed will eventually start to spread on their own.

Having explained Fisher's method, Bonython suggested to John Murchison it might be useful to their shared political cause. He lamented 'the continual growth of government control, of denial of the right to have private property', which, he said, led to 'Big Brother' systems and 'eventually dictatorship'. He then informed Murchison he had passed his address on to Fisher and suggested the pair meet in person.

'He is not in it just to earn his own living – but to help the cause,' Bonython said. 'I mean the cause of freedom, the maximising of choice, a market economy, as opposed to controls.'

According to Dr Jeremy Walker, a historian of political economy and earth sciences at University of Technology Sydney, Bonython's letter is 'one very significant document among many' outlining the start of a 'permanent corporate influence and disinformation campaign' in Australia. Walker, who found the letter in an archive maintained by the Hoover Institution in the US, says that not only does it demonstrate a clear awareness among industry figures like Bonython of how unpopular they were with the public, it is also a clear statement of intent. Bonython wanted to change the way Australians thought about the world, and to do that, he felt it was necessary to shape how ideas were presented to both the public and policymakers – without revealing the origin of these ideas.

'The point is to just saturate every aspect of the public sphere with messaging that supports their corporate interests without disclosing the origins of that messaging,' Walker says. 'Its purpose isn't to be true, or even coherent; its purpose is to be repeated often.'

Fisher's think tanks wouldn't be the first – the Institute of Public Affairs had set up shop more than 30 years earlier, back in 1943, with financial support from Keith Murdoch and mining interests, among others. Fisher's innovation was to connect different think tanks operating in different countries so they could coordinate internationally and turn them into a production line.

Using his thick Rolodex of contacts among the rich and wealthy, Fisher had already set about expanding beyond Britain, often through proxies or in collaboration with others. According to documents discovered by Walker, Fisher's own notes describe how he had been involved with 40 institutes in 21 countries by 1987.

The first think tank outside of the UK he was involved with was the Fraser Institute in Vancouver, where he was invited to be a co-director in 1974. This was followed by his visits to Australia in 1976, where he met with business figures including John Bonython, Hugh Morgan and stockbroker Maurice Newman and became aware of a Sydney-based maths teacher named Greg Lindsay. Lindsay had already set up the Centre for Independent Studies in 1976 in his shed, and, though Fisher was cautious, documents obtained by Walker show Fisher was advising Lindsay on fundraising strategy.

Similar records show how in 1977, Fisher and a future director of the CIA, Bill Casey, set up the International Center for Economic Policy Studies in New York (later renamed the Manhattan Institute); the same year, Fisher co-founded the Adam Smith Institute in London. By 1979, he was living in San Francisco, where he founded the Pacific Institute. Crucially, Fisher founded the Atlas Economic Research Foundation in 1981, which functioned as an umbrella organisation to coordinate logistics and funding between the different groups. This organisation would later be rebranded the 'Atlas Network', something akin to a trade name.

Unlike the individual think tanks, Walker says, the Atlas Network rarely publishes anything under its own banner and is not publicly involved in any specific campaign. Its role is to allocate resources and train personnel before shipping them out to wherever they are needed around the world.

'What you end up with is a heap of different sources with similar messages that look like they came from different places but all trace back to the network,' Walker says.

From the beginning, he says, these organisations were awash in oil money. The prototype, the Institute of Economic Affairs, was founded in 1955, and Walker says it enjoyed ongoing funding from Shell and BP from the early 1960s on. In 1979, six companies gave the Centre for Independent Studies $5000 a year for five years: Shell, a corporate precursor to Rio Tinto, BHP, Western Mining

Corporation, Santos and *The Advertiser.* John Bonython, meanwhile, later joined the board, becoming the first Chairman of the Board of Trustees. Since 1984, the CIS has run a lecture series named for John Bonython.

Early on, their main interest was trade liberalisation and the free market, subjects which appealed to the wealthy businessmen whom Fisher courted. Later, that would change, thanks in part to a fateful visit to the US, where Fisher met with Charles and David Koch.

When Fred Koch died in 1967, his four sons inherited a majority share in an oil company, granting them an instant fortune. In 1983, Charles and David Koch bought out their brothers and became majority owners themselves.

Theirs had always been a political household. Their father had co-founded the John Birch Society, but the racist conspiracy-mongering of the Birchers didn't really fly with Charles Koch. His drug of choice was supplied by Friedrich Hayek and Ludwig von Mises, free market economists whose ideas radicalised a generation that would later declare 'taxation was theft'. Charles flirted with right-wing organisations, but it was after meeting Fisher that he learned the 'method' and used it to fund their own think tank, the Cato Institute, in 1977. This would then be followed by several others, including the Heartland Institute. With time, the management of Fisher's network and the think tanks run by Charles Koch would begin to overlap. Upon Fisher's death in 1988, Koch became the dominant influence on the board of the Atlas Economic Research Foundation, vastly expanding his reach.

'And it's exactly at that point they start doing anti-climate policy,' Dr Walker says.

The changing of the guard within the Atlas Network just happened to coincide with the creation of the United Nations Intergovernmental Panel on Climate Change and the writing of the internal Exxon memo outlining its intent to publicly emphasise doubts about the emerging science. The next year, Fred Singer, an Atlas Network alumnus, published a book that would become the template for those denying climate change, *Global Climate Change: Human and natural influences.*

Singer wasn't just a climate denier, he was *the* model for climate denial. A physicist, he had started out as a straight-down-the-line scientist and environmentalist but broke bad after being denied a promotion with the Environmental Protection Agency. His first forays into the bullshit economy included fuelling doubts about the science of second-hand smoke, the hole in the ozone and acid rain. But his first organised denial group involved boosting for Reverend Sun Myung Moon, a Korean billionaire with a messiah complex who claimed to be on speaking terms with God, Moses and Jesus. When the Reagan era ended – the former US President was a big fan – Moon's influence ended too. Singer needed a new hustle and, willing to take money from just about anyone, he switched focus to climate change. He had penned columns for Moon's newspaper, *The Washington Times,* with titles like 'Chilling Out on Warming' and with oodles of industry money being sunk into generating doubt over the science, the pivot wasn't difficult. With the book out, Singer did the rounds on the speaking circuit.

As this was all getting going, the AIP had laid down its official position on climate change and APEA began haranguing governments over the issue. The power of this early campaign can be measured by its results. Climate change was replacing the hole in the ozone layer as the next big international challenge, and in 1989 the Hawke government was starting to think about action.

A cabinet document from that year, marked 'cabinet in confidence' but now freely available to the public through the National Archives, describes how the government considered human activity, particularly 'population growth, burning fossil fuels, agriculture and deforestation' to be the cause of 'growing concentrations of CO_2 and other "greenhouse gases" (oxides of nitrogen, CFCs, methane) in the atmosphere'. To address this problem, the Hawke government wanted to match similar US research programs by resourcing its own and running a public awareness campaign to drum up support for its efforts.

From the beginning, the Hawke government encountered

friction from within the bureaucracy. In an appendix surveying the views of various departments, the Department of Foreign Affairs and Trade carefully gave its support, even as it sternly warned that the 'economic and trade implications of the greenhouse effect are so significant that they must be faced squarely'. It also advised caution about the wording of any press release that might expose the government to criticism due to the inconvenient truth that Australia was the world's largest coal exporter. The Department of Primary Industries and Energy – sometimes referred to derisively by climate scientists of the period as 'DoPIE' – resented anyone stepping on its turf. The department pointed to all the research it was already doing on climate change through the Bureau of Rural Resources – a forerunner to Australian Bureau of Agricultural and Resource Economics ABARE – and the Bureau of Mineral Resources and felt it should decide policy on the issue, not those hippies over in the Environment Department. Meanwhile, Treasury – likely responding to the influence of penny-pinching finance minister and climate denier Senator Peter Walsh from Western Australia – raised concerns about 'a risk of wasting resources in funding additional programs given the uncertainties on basic facts available, and forthcoming from existing research'.

'Australia already spends almost $14 million annually on "greenhouse effect" related research, and the bulk of funding and research on the greenhouse effect is inevitably done by larger countries,' it said.

It concluded: 'Treasury considers that any increased work on this issue should be done by reallocating resources from other less pressing areas.'

This dour response foreshadowed the reactionary role these departments would play in putting a handbrake on climate action over the next decade. It was like a little insurgency within the bureaucracy. At first pro-business guerrillas were operating in the metaphorical hills, but with time they grew bolder. As the fighting spread to the cities, successive governments would throw up their hands in surrender.

A watching brief on climate change kept by the Department of the Prime Minister and Cabinet stands as a remarkable relic from the end of the 1980s. A copy of the briefing folder's contents reveals old cablegrams from Australia's delegates overseas asking for instructions from the leadership, copies of speeches about climate change, policy outlines, a flyer for the world coal conference – and a single newspaper article headlined: 'Hot air and the cold facts'. The article was a dispatch from Bryan Boswell, reporting from Washington for *The Australian* about the 'so-called greenhouse effect'. Whoever added it to the brief had carefully underlined certain sections in blue biro:

> The report from <u>NASA</u> analysts Roy Spencer of the Marshall Space Flight Centre in Alabama and John Christy of the University of Alabama, said that <u>examination of 10 years of measurements by weather satellites had failed to find any evidence at all of global warming from 'the greenhouse effect'</u>. They suggested that <u>violent rises in temperature quoted</u> by other scientists were <u>part of the normal weather pattern.</u>
>
> [...]
>
> Hurriedly putting themselves on the fence in the debate, they both added that their data was 'tantalising', but that it would take <u>another decade of measurements to be certain.</u>
>
> [...]
>
> That report, released this year, was aimed at the effect of the so-called acid rain – also blamed on CO_2 emissions – but confirms that <u>only 2 per cent of US lakes are actually acidic and that nearly 70 per cent of those were acid in pre-industrial times.</u>
>
> <u>Other studies seem to show the polar caps are growing, not shrinking</u>, as 'global warming' would suggest.
>
> [...]
>
> Meanwhile Hugh Ellsaesser of the <u>Lawrence Livermore Laboratories has produced data showing that most of this century's warming took place before 1938 – well ahead of the post-World War II rise in carbon dioxide concentration.</u> From 1938 to 1970, he points out, temperatures plunged so sharply that a new ice age was being widely forecast.

It would have been a powerful blow to momentum. NASA scientist James Hansen had just raised the alarm about climate change when he testified before US congress, but here was Spencer, another NASA scientist voicing a dissenting opinion. The implication was clear. If even NASA couldn't agree on what was going on, why should the Australian government be anxious to act?

A cursory read of the article offered no biographical detail about these men. Spencer was a believer in intelligent design; both he and Christy would later become experts attached to the Heartland Institute; Ellsaesser sat on the science advisory board for the now-defunct George C. Marshall Institute, which was later described by a *Newsweek* cover story as a 'central cog in the denial machine'. Like the scientists who ran interference for the tobacco industry before them, these men had appointed themselves the arbiters of 'good science'.

They were the merchants of doubt, and their opinions were in print, highlighted and ready to be slipped in front of decision-makers at the highest levels, right as the Australian government began considering action against climate change.

A few years on, the worldwide disinformation machine was going strong in Australia. Fred Singer contributed an essay to an IPA pamphlet in 1991 declaring there was no scientific consensus on climate change and followed this up with a tour of Australia the next year; Bob Foster, BHP's man who had so desperately wanted to drill the Great Barrier Reef, had been out claiming there was 'no evidence' that burning fossil fuels was driving climate change since 1988; and newspapers ran thought bubble comments from people who blamed global heating on sun spots or the tilting of the earth's axis. Climate denial guides were printed complete with 'conversation trees' with predicted questions and suggested answers, showing how to entrap anyone arguing that humanity was responsible. The arrival of the internet in the mid-1990s created a frictionless environment for the sharing of information, and there was no man in a better position to take advantage of that than Hugh Matheson Morgan.

Morgan was a rich man on a mission. From the early 1980s, he had been a stalwart figure in right-wing politics. He served as CEO of goldminer Western Mining Corporation, would be appointed to the board of the Reserve Bank of Australia for the first time in 1981 and then again in 1996. All the while he kept close ties to the US. Later, Morgan enjoyed unparalleled access to John Howard, boasting the kind of relationship where he could pick up the phone and get a meeting on the same day. Thanks to this pull, Morgan would later claim credit as the man responsible for keeping Howard from ratifying the Kyoto Protocol.

Following the pattern set by the Koch brothers and Fisher, Morgan had busied himself through the late 1980s and early 1990s forming his own set of single-issue advocacy groups. His chief agent in this task was a man named Ray Evans.

After funding cuts at Deakin University cost Evans his job, Morgan took him in and the electrical engineer would spend the next 20 years of his working life as WMC's 'corporate theologian'. As La Trobe University academic Dominic Kelly put it: where Morgan was the public face, Evans was the speechwriter. Serving as Morgan's proxy, Evans made a name for himself as a shouty right-wing activist for whom accusations were confessions, the standard-bearer for the 'political troglodytes and economic lunatics' who plagued Bob Hawke's government. He started his political life in an anti-communist Labor Party group at university before switching teams and enlisting as a foot soldier in the culture wars. As a committed Christian, he helped carve the 'New Right' from the whitest of marble.

Evans was quite possibly Australia's original edgelord. The internet may have made trolling a political art form, but Evans understood the power of rage-farming earlier than most. His was a tactical racism. If reporters were suckers for scandal, he pioneered the strategy of saying something outrageous to the press, knowing it would be picked up and amplified when the political left responded with fury. Whether he actually believed what he was saying is beside the point. Every stupid, racist, misogynist or classist thing he said helped to heave the acceptable boundaries of political debate rightwards another inch.

And there was no satisfying Evans. Allies who didn't 'go all the way' were wimps; his enemies were given no quarter. In this, he was a man built perfectly for the era of infotainment. Where others believed in the superiority of fact and reason, Evans understood his audiences wanted to be entertained. They didn't care about what the IPCC said or the results from global average temperature series; once an enemy was identified, they wanted the catharsis of having overcome the other. There was no way to negotiate with or appease Evans. The choice he offered those who opposed him was simple: surrender or be crushed.

Together, Morgan and Evans founded the H.R. Nicholls Society to encourage union busting, the Samuel Griffith Society to defend the federal principles of the constitution, the Bennelong Society to counter the Indigenous land rights movement and, crucially, the Lavoisier Group to aggressively attack the suggestion that humanity had any role at all in causing climate change.

The Lavoisier Group materialised after Australia signed the Kyoto Protocol. It was named for French chemist Antoine Lavoisier, the father of modern chemistry who discovered oxygen, hydrogen and the physical mechanics of combustion. Over the course of his life, he married a 13-year-old girl and was guillotined by revolutionaries – an untimely demise at the hands of the left which made Lavoisier a martyr to Evans and his colleagues.

Their inaugural conference was held in May 2000. Morgan himself gave a speech attacking then-federal environment minister Robert Hill. Knowing that a little good, bad press could help build a public profile, Morgan also took the liberty of describing Australian Greenhouse Office statements on climate change as 'Mein Kampf-declarations'. The official founding took place a few months later, at the insistence of Peter Welsh, who wanted a countervailing pressure group to run interference against efforts to address climate change.

On the board were Walsh, the former Labor finance minister; Ian Webber, who held directorships with Western Mining Corporation director and Santos; Harold Clough, the Western Australian multimillionaire, Liberal Party financier and IPA board member; Peter Murray, a coal industry consultant; and Bob Foster, prominent climate denier and petroleum engineer. Ray Evans served as their fearless leader, with Morgan and former Nationals senator John Stone taking their turns later. In sum, Lavoisier was an oligarchs' club.

There were other groups in the mix too – in June 1990, alongside the IEA and CIS, the Tasman Institute had begun operating as a free market think tank with financial backing from banks, miners, department store chains and Rupert Murdoch. From the moment it started up, the organisation's researchers went to work attacking the idea of a carbon tax. The creation of these institutions and organisations would be replicated all the way through the mid-2000s and beyond using much the same tactics. Figures like South Australian Liberal Senator Cory Bernardi and One Nation Senator Malcolm Roberts would forge their own connections to The Heartland Institute in the US, today backed by billionaire hedge-fund manager Robert Mercer. When they returned home reborn, they would dedicate themselves to telling Australians they'd been lied to about climate change, that it was all an effete conspiracy by craven elites and the UN.

Lavoisier, however, was a different beast. Money, connections and hostility made the group effective in the heady period of the late 1990s to early 2000s. The group might have been too aggressive to have much direct influence on policy, but their strategy was always to play the man, not the ball. They considered themselves rabble-rousers, hellraisers – the barbarians at the gate.

Morgan would brag that the literature on the group's website received a huge number of downloads, but there was no way to know what he considered 'huge'. Primarily, the group acted as a steering committee for all the other organisations controlled by its individual members. Its membership could publish all the books and pamphlets and bluster their way through as many speaking events as they wanted, but it was the long reach of their personal networks that gave them oomph.

The collective wealth controlled by its members ran into the billions, and their positions overseeing large companies or groups of companies – often operating in the oil, gas and coal industries or their supply chains – meant their particular worldview had a way of trickling down. It would not have been likely, for instance, for an ambitious young executive to rise through the ranks at Clough Industries without at least appearing to share the outlook of its owner, Harold Clough. Neither would it have been possible for an ambitious young MP to hit him up for a campaign donation – as Tony Abbott would when he was running the campaign that ultimately put Pauline Hanson in jail – if they weren't willing to return the favour down the line.

It helped, too, that Evans had linked up with similar outfits in the US. The most significant of these connections was the Competitive Enterprise Institute in Washington, a group which, according to the Climate Investigations Center, received US$2.1 million from ExxonMobil between 1998 and 2006. Thanks to this international connection, Lavoisier acted as a portal through which to channel the latest talking points into Australia from the US.

Bringing home these Big American ideas, Evans wanted to put them to work in Australia. He would do so with Lavoisier's self-styled 'free thinkers', who operated with a fundamentalist zeal. A favourite insult was to describe climate science as a religion and those who worked in the field as its high priests, perpetrating a scam to get access to grant funding. It was nonsense, of course. The oil industry itself had supported much of the early atmospheric research that confirmed the existence of the greenhouse effect, and it was still monitoring the situation. In 2002, Exxon's Australian subsidiary, Esso Australia, even sponsored a joint University of Sydney and University of Newcastle climate monitoring project on its offshore platforms.

Exxon's own researchers had long since established the basic mechanics by which the greenhouse effect worked. It wasn't rocket science, but it was an explanation already too technical for most people who weren't interested in retaking high school physics. But Evans, whose livelihood was bound up in the resources sector, would prove ruthless in exploiting this basic knowledge gap among the general public and decision-makers. Like a pro-business Mao or an anti-tax Malcolm X, Evans was ever ready to storm the capital in defence of his god-given right to emit as much carbon dioxide as he could generate.

11

The Penny Is Going to Drop

It was 2009, Lavoisier's star had begun to fade and the global financial crisis had left the world economy a smoking ruin. Australia was a rare exception: its coffers were about to burst from a spike in the iron ore price on the back of a Chinese construction frenzy. With tradies pulling $120K-a-year salaries, climate change seemed almost academic to the country's movers and shakers. To many, it was one of those distant problems, somewhere far over the horizon, an interesting issue, perhaps, but one for someone else to deal with.

There was at least one group of people, however, who cared very passionately about climate change at this time. The issue had come onto the Sydney Institute's radar the moment Kevin Rudd tried to do something about it. He had taken office in 2007, ratified the Kyoto Protocol and started talking big about Howard's old plan for an emissions trading scheme. This attention to the issue would continue throughout the prime ministership of Rudd's successor, Julia Gillard.

The Sydney Institute was a rough equivalent to the Atlas-aligned Manhattan Institute, a policy forum that cleaved to the political right. The organisation's sudden interest in climate change coincided with a flurry of media interest in reflexively contrarian views disputing the scientific consensus that the planet was getting hotter. Figures who rejected that consensus would become regulars at Institute events, giving talks and presentations with titles like 'The Theology of Climate Change'.

Against this backdrop, Ian Dunlop, a petroleum engineer and 35-year veteran of the oil industry, was fighting to get on the Institute's program. Dunlop had been lobbying the Institute's executive director, Gerard Henderson, for some time to give his own talk on the ramifications of climate change. Unlike the climate deniers, who had no trouble getting stage time at Institute events, Dunlop was deeply concerned about the risks posed by climate change. He had contributed a foreword to a book titled *Climate Code Red* published the year before, and he wanted to provide balance to those who dismissed the science.

Rather than a standalone presentation, however, Henderson – a former political adviser to John Howard – asked Dunlop if he would participate in a debate on the climate change 'question'. His opponent would be Ray Evans.

The prospect of a showdown drew so much interest from the Institute's membership that they moved the venue to the Metcalfe Auditorium at the State Library of New South Wales at the last minute. As the audience settled into the red, folding lecture-hall chairs in the wood-panelled room, Henderson set up the session by comparing it to earlier debates about the existence of god. He introduced Evans as a former Deakin University engineering academic, a 20-year veteran of Western Mining Corporation, secretary of the Lavoisier Group and author of an ironically titled pamphlet, *Thank God for Carbon*. Henderson helpfully informed his audience that people could pick up a copy from the book table at the back of the room.

He then turned to introduce Dunlop by rattling off his CV: Dunlop had served as the head of the Australian Coal Association, CEO of the Institute of Company Directors and chairman of a group of experts who, in 1998, had developed the first proposal for an emissions trading scheme under the auspices of Australian Greenhouse Office. What Henderson did not explain was that Dunlop had first learned about climate change while working on long-range planning projects with Shell beginning in the late 1960s. He had been taking the issue seriously for most of his professional life, even when those back at company headquarters refused to accept the reality. He had enjoyed his time in the industry right up until 1991, when he broke ranks, partly due to the company's climate response.

Evans came out swinging, opening with an attack on the Labor government for wanting to set up an emissions trading scheme on the basis of their belief, as he described it, 'that by controlling anthropogenic emissions of carbon dioxide, in effect by turning off the CO_2 tap, they can control the world's climate'.

'That belief has no more foundation than the belief that dancing in drought times around, say, the lawns of Parliament House in Canberra, will bring forth rain,' he said.

As far as Evans was concerned, all the CO_2 being pumped into the atmosphere was just progress. In fact, he claimed, there was 'nothing in the climate record – nothing, zero' to suggest that atmospheric carbon dioxide had 'any measurable impact on the Earth's climate'.

His was a belief system rooted in fatalism. If the climate was changing – which Evans was never totally willing to concede – he felt that humanity should just accept it was powerless. Having said so, and with 15 minutes left to range free, he then embarked on a breathless Gish gallop – a rhetorical trick commonly employed by evangelical preachers, in which a series of arguments are made in quick succession. They may rely on half-truths, outright lies or misrepresentations – it doesn't really matter. It only matters that this blizzard of nonsense is delivered thick and fast, overwhelming their opponent, obliging them to respond, when they take the floor, to every single claim in turn. Because it is easier to lie than it is to correct, working through each falsehood chews through precious time, undermining the opponent's credibility in the eyes of an audience unfamiliar with the tactic.

Those who cared about climate change, Evans charged, were engaged in a form of 'religion', which he described as a 'millenarian frenzy', led by 'prophets' who were 'linked to radical ideologies'. It was a rare insight into the psychology of a man aggressively opposed to the notion that humanity had any influence on the climate. He was essentially playing the role of inquisitor stamping out heresy before an audience that fancied themselves free thinkers.

Having allowed Evans to set the terms of the exchange, Henderson invited Dunlop to make his rebuttal. Dunlop had flown in from Europe the day before and had laryngitis, which didn't help. Still, he attempted to diffuse the tension created by Evan's scorched-earth opening with a joke.

'After Ray's comments, I feel hugely underdressed,' Dunlop said. 'I didn't bring my religious habit.'

He then presented the audience with a methodical review of the best available science at the time. Climate change was almost certainly caused by human activity, he said – and if it wasn't, we should be even more concerned, as that would mean humanity couldn't control what was coming. The best way to proceed, he suggested, was to treat it as an exercise in risk management. According to the precautionary principle, it is better to act now, rather than dawdling only to find later you have wasted precious time and everything is considerably worse than it might have been.

Following the presentations, attendees were invited to ask questions. Almost instantly, audience members accused Dunlop of having 'failed' to prove the connection between CO_2 levels and climate change, as if he were a one-man IPCC. Others wondered why it was even worth attempting to do anything at all, asking, 'Do we really want to see our economy go down?' In response to a rare question directed at Evans, about the physical reality that easily accessible supplies of oil, gas and coal would eventually run out, he answered by paraphrasing former Saudi oil minister Sheikh Zaki Yamani: 'The Stone Age didn't come to an end because they ran out of stones.'

Without missing a beat, Dunlop pointed out the obvious sleight of hand.

'If I could just add to that,' he said. 'Ray, I think you should add the second part of the quote, which is *the oil age will end long before we run out of oil*. And that's what's happening.'

Evans might have been caught out cherry-picking quotes to suit his agenda, but he powered right on through. 'Coal is so cheap, so plentiful in Australia,' he said, that there was no reason to stop burning it for power. He grew particularly passionate when attacking climate modelling as 'worthless'.

These comments were red meat to this audience. Dunlop's wife and a friend had attended that night to support him, but when those sitting nearby learned of their relationship, they turned their barbs on his guests for being associated with a heretic.

Evans predictably used his closing statement to charge his political enemies once again with religious frenzy, but Dunlop took wider aim. There was an audible frustration in his voice as he countered: 'In this country, the dominant vested interests and the power structures have been built up around the fossil fuel industries.'

'These are the people who have the money, the influence, the ideological grunt, if you like,' he said. 'To suggest to them that after fifty or a hundred years of that evolution, you suddenly have to change the way you think, transform into a completely different mindset – that's not easy.

'And that is why everybody is out there defending the status quo.'

It was an extraordinary statement, coming from a former petroleum engineer. To emphasise his point, Dunlop quoted an observation that he attributed to German philosopher Arthur Schopenhauer: 'Truth goes through three phases. First, it's ridiculed. Second, it's violently opposed. And third, it becomes self-evident.'

'I would argue we're just coming out of the second phase, and very soon the penny is going to drop,' Dunlop said.

Considerable research has gone into trying to understand the psychology of climate denial: how climate deniers think and why the broader public hasn't mobilised to demand action as awareness about the risk posed by climate change has spread. Among deniers, logic traps and fallacies abound, as does a general trend towards conspiratorial thinking. Some people have been genuinely bamboozled by the global disinformation campaign seeking to obscure the reality of climate change; others just don't want to confront the threat climate change poses to their normal, everyday lives. A subset of that group are knee-jerk contrarians who treat it as a basic political proposition: if those on the 'other team' care so much about climate change, the idea must be met with total opposition.

Where there hasn't been as much research, however, is in efforts to understand the psychology of those who continue to work for fossil fuel producers.

According to the International Energy Agency, some 11.8 million people were employed in the oil and gas industry worldwide as of 2019, with 1.1 million of them working across Oceania and the Asia-Pacific. What this meant was stark. For every person who chose to quit the industry over concerns about climate, there were thousands more who stayed.

These statistics raise a startlingly simple question: why aren't more people quitting oil companies? At this point in human history, there is a wealth of information showing that the extraction and burning of fossil fuels harms the climate. Humanity was already pumping 37.12 billion tonnes of CO_2 into the atmosphere by 2021, and these figures were easy to check, thanks to the internet. Every other week, there seems to be a new story about how oil, gas and coal companies have actively tried to hide the problem to extend the life of their businesses or lobbied to stall action. So why stay?

The answer is obvious: it's a living. Finding a new job is hard, and people have mortgages to pay. But that just prompts more questions. Having confronted the reality of climate change, how do people rationalise their decision to stay? How do they keep going, knowing that they are helping to create a frightening future for themselves and their children? How do they convince themselves that, at their core, they remain a good person?

There is research suggesting that the human brain is just not equipped to handle the implications of climate change, which is why people don't make changes in their own lives to help tackle global heating. But such studies tend to reduce the problem to a question of individual action, blaming the victim, not the perpetrators.

This 'victim blaming' is a strategy the oil companies themselves pioneered in their public relations campaigns. The idea of the 'carbon footprint' – originally invented by sustainability advocates – was hijacked by BP in 2003 to redirect blame from its role in causing the problem to begin with. Ever since, it has been used to make people feel guilty about their personal contribution to global heating. In truth, the contribution of the individual is greatly outweighed by the institution. According to a peer-reviewed study by Richard Heede of the Climate Accountability Institute published in 2013, just 90 of the world's largest fossil fuel and cement producers were responsible for nearly two-thirds of carbon dioxide and methane emissions released since the Industrial Revolution. Just under a third of all these emissions could even be attributed to just 20 individual companies.

This is what makes oil, gas and coal workers a unique group among the general population. From the lowliest roughneck to the highest executive, they are subject to the same pressures as everyone else in the wider community – but as a group, their proximity to the process of extraction, transportation, refinement and burning of fossil fuels means their individual and collective decisions have greater resonance.

None of this would be an issue if major petroleum companies had a credible plan to shift their business away from fossil fuel extraction and production. If they did, their employees could clock in knowing they were actively working towards a clear, tangible goal. That is not the case, however, as almost every oil and gas producer in the world remains rigidly committed to 'core business'. BP is emblematic of this tendency. Having previously abandoned its 'Beyond Petroleum' rebrand, BP recommitted itself to substantial new oil investment after prices spiked following the Russian invasion of Ukraine. The money was simply too good.

A psychologist might say that the employees of fossil fuel companies experience 'cognitive dissonance': feelings of discomfort or anxiety that arise from inconsistencies in a person's ideas or beliefs. If you believe that burning fossil fuels causes climate change and that it poses a serious threat to human existence, it's much harder to maintain the belief that you're a good person while you keep going to work for an oil company every day. Old-fashioned denial is another psychological defence mechanism people use to cope with this anxiety, refusing to recognise the reality of a situation to avoid the uncomfortable feelings of dissonance.

These concepts may partially help answer why more people don't quit, but not completely. A more complete answer means looking further afield to other theories. In the 1950s, two social scientists, David Matza and Gresham Sykes, developed the theory of neutralisation to explain how those who did bad things 'neutralised' or justified their actions to themselves so they did not lose their fundamental belief that they were good people. Matza and Sykes identified five techniques: *denial of responsibility* (pointing to mitigating factors); *denial of injury* (insisting that the harm inflicted was not as bad the victim claims or that the harmful action was actually beneficial); *denial of victims* (claiming there was no victim or that the victim deserved it); *condemnation of the condemners* (attacking those who call out the bad behaviour); and *appeal to higher loyalties* (claiming there is a higher power, above the law, that required the harm to be committed).

Originally developed to analyse juvenile delinquency, these neutralisation techniques have been used to understand the actions of gang members, organised criminals and genocidaires. Professor Rob White, a criminologist at the University of Tasmania, says these neutralisation techniques could just as well be applied to fossil fuel workers trying to justify their ongoing work. He also cautions that while the idea of neutralisation is a 'useful starting point', it is not without its limits.

'I also think some people just don't give a shit,' White says. 'Even that expression, *don't give a shit*, implies a position of power, a position of material benefit that overrides everything else.'

Others have approached the question from another direction. Dr Grace Augustine of the University of Bath and Dr Birthe Soppe of the University of Innsbruck and University of Oslo set out to investigate the phenomenon of 'climate quitting' – people who were leaving the oil and gas sector over concerns about climate change. Between July 2022 and March 2023, they spoke to 32 current and former oil industry workers from every inhabited continent. It was a small group, not a large enough sample from which draw any firm conclusions, but there were enough similarities between the stories they heard to piece together something resembling a decision tree – that is, a map of the choices available to a person when confronted by a situation that takes into account potential outcomes, consequences and costs.

What they learned was something often overlooked by activists and analysts alike: people really liked working for oil companies. These jobs followed a boom-and-bust cycle, but in a precarious world, the good times – when they came – were very good. The work was hard and complex, but intellectually satisfying. People were given real budgets and trusted to tinker. For those in developing countries and working-class communities in the developed world, these jobs often represented a clear pathway to upward social mobility. When their sons and daughters went to work for an oil company, it meant they had a shot at a comfortable life. This made these jobs prestigious, especially in historical oil-producing regions like Houston, Aberdeen or Alberta. Many felt they were actively contributing to their communities by 'helping keep the lights on'. When their company sponsored the local sports team, there was genuine pride. Most importantly, they were well paid.

'The industry has such a social legacy, and of course they fuel that rhetoric,' Dr Soppe says. 'This kind of legacy-identity the industry tried to coin is very much reflected, of course, in people's considerations when they join.'

When researchers asked people how their nine-to-five work contributed to ecological harm, their first instinct was often to rationalise the problem away. A person might have kids to feed and a mortgage to pay; they might see the company as their ticket out of a bad situation; or they might think climate change was a distant problem that someone else would get a handle on sooner or later. There were other, more specious rationalisations, too. Some reasoned that if they weren't doing this job, someone else would, so they might as well take the money. Others would look around and take comfort in the knowledge they didn't work for the 'bad-bad guys'.

'At least we're not working for Exxon,' people joked.

There were even those who simply had a naïve faith in what the company was telling them. When the company promised an impressionable graduate that it had dedicated itself to change, that a transition was inevitable and they could be part of the solution, they accepted it. Eventually, often after a period of years, there came a moment of clarity when they stopped and realised that the 'inevitable transition' they were promised had never arrived.

For all the pressure social movements were bringing to bear on oil company executives from the outside, those working inside the glass towers told the researchers that they'd never felt any pressing need to quit. At first, the heightened scrutiny of their industry made them more aware of climate change, but their response was to attempt to create change from within. In one way or another, they had themselves moved onto committees or projects that might help to reduce emissions or worked up proposals for new business lines in renewable energy. Often, they would pour themselves into these projects, putting together detailed presentations and analysis to justify the investment risk. And the companies often welcomed these efforts. They needed them. The presence of good people who were trying to do the right thing allowed those doing nothing to continue to feel like good people, too.

It was only when their efforts failed to get any traction or were actively shut down by management that something changed for those good people helping to salve others' consciences. Sooner or later, resentment stirred. As they looked around their workplaces, they started to feel trapped. Worse, some reported experiencing a growing sense that they were helping build a dystopian future every time they clocked in. Some interviewees went so far as to compare it to working as a stormtrooper on the *Death Star*. One medical professional described feeling like 'a doctor working in a hospital run by Hitler', saying their 'degree had been hijacked' by oil and gas.

Yet it still took something more to finally shake them loose – internal politics, a dumb decision by headquarters, a restructure, a jerk boss, witnessing something illegal. Even then, despite it all, some chose to stay – though they felt bad about it. It was the engineering equivalent of 'shooting and crying' – an expression coined by Israeli scholars for when soldiers depicted in film are shown feeling remorse for their actions during service. In the moment, they had all done the bad thing together – no-one refused or intervened to stop it. Whatever they felt about it at the time, they'd carried out their orders like good soldiers. Afterwards, they all felt bad, and this display of remorse relieved them of guilt.

Those who remained with the oil company were engaged in something similar. In effect, they were drilling and crying.

In fairness, leaving employment genuinely wasn't possible for some. Maybe they worked in a developing-world country where there really was no option, or in their region there was nowhere to make an obvious exit. In my own research, I spoke to oil company employees who, feeling like they had been forced to stay put, turned to sabotage, one $10,000, productivity-killing mistake at a time. Others chose to leak to reporters, or were desperate to be fired. Because of the generous redundancy packages in some workplaces, it was understood that at any given time a percentage of the workforce was angling to be laid off by deliberately underperforming. In one story I heard, an employee wore an Extinction Rebellion t-shirt to a meeting in hopes of being fired but in doing so overplayed his hand with management, who saw through the provocation.

Those who left faced other struggles. Many reported first having to overcome a powerful feeling of anxiety. All their friends might have worked at the company, which meant they were leaving behind a social support network. Looking for a new job was also hard. If oil, gas and coal were truly the new tobacco, some wondered whether anyone would actually hire them with a decade at a fossil fuel company on their CV. Despite these setbacks, many did find new work. According to Augustine and Soppe, those who made a conscious choice to leave over concerns about climate change often landed at renewable energy companies, activist organisations or policy groups working on the transition, depending on what was happening where they lived. Their research helped outline the choices faced by those grappling with inner conflict about whether they were doing the right thing, but to take the first step towards quitting required an initial moment of clarity or insight. For all those who quit, there were thousands more who did not.

To find out more, I asked Ian Dunlop why he thought more people weren't making the switch. After all, he had done so early and with stunning success. Those who worked within the oil industry were smart people, he said. They were good scientists, engineers and analysts who had genuinely dedicated themselves to their work and were striving to the best in their field.

'The problem with the fossil fuel industry is that you get people who have committed themselves for a number of years and who have specifically committed themselves to excellence,' Dunlop said. 'It's extremely hard for them to accept the fact this is happening, that their commitment to excellence in one arena is causing harm in another.'

The penny dropped for Joshua about three years into a four-year stint in the oil industry. His university required him to do a twelve-week professional work placement to graduate, and the only real opportunities for someone with his particular skillset were in mining, petroleum or government. He had no particular preference at the time, but ConocoPhillips just happened to be the first to return his calls. Before he knew it, they had flown him out to Darwin, put him up in a hotel and sent him to work. They even paid him.

Darwin was fun for the farm boy from rural Australia, and for a brief moment it offered a taste of the good life. Compared to hard rock mining – the process of ripping minerals from the earth – work in oil and gas was more complex, more exciting and way better paid. In his program, the oil companies only took the best, so making it through the door was a mark of distinction. When the American oil company later invited him into their three-year graduate program, it was a signal they had considerable faith in him.

Joshua was a pretty big outlier among his peers in the graduate program. Unlike the others, he'd always had a natural cynicism towards power and the powerful. More to the point, he wasn't interested in making the company his life. He had a skill, he liked using it, and he was good at his job, but he wasn't about to kill himself to get promoted to management, where he'd no longer be doing the work he enjoyed. The company still turned it on for him, though, as it did for all of the new graduates, moving them around to different departments so they could get a sense of what they enjoyed most. Once, as the east coast burned during the Black Summer bushfires, the company flew the graduates business class to Houston, Texas, to meet others in similar programs from around the world.

'It's a way of training up people to indoctrinate them into how the company works,' Joshua says. 'Companies have a way of – there's a ConocoPhillips way of doing things. It's corporate speak, it's about making you aware of their values and their mission and how that all works.'

Around that time, Joshua started thinking about climate change. It was hard not to. There were viral videos of walls of flame in the Blue Mountains leaping vertically up cliffs, and the newspapers delivered to the office reported the latest events in detail. Curiously, he thought, none of these reports seemed to connect these events with a changing climate, or the fossil fuel companies' role in making it happen. He tried bringing it up with his colleagues, but it was like talking to the wind. No-one he worked with seemed interested.

'It was unspoken,' he says. 'I always felt very keenly that there were a lot of people who were either deniers or would judge you very, very harshly for bringing something up along the lines of climate or environmentalism. Particularly the older people.

'You didn't speak about it. It was taboo.'

If Joshua thought the old-timers at ConocoPhillips were brushing him off, though, things would only grow worse. As the fires burned, news broke that the American firm was selling off its Western Australian and Northern Territory divisions. Santos would drop $1.39 billion to pick them up, and an unsettling feeling washed over the workforce. Among Joshua's older colleagues, Santos had a bad reputation. Few were thrilled by the development.

Joshua had been starting to look more closely at getting himself on internal projects working on emissions reduction. On his own time, he began familiarising himself with Santos' annual climate reports, which tracked how the company was performing against the targets set by the Paris Agreement. These reports were based on data supplied by the International Energy Agency, which had historically been an oil-friendly institution.

That changed in May 2021, when the agency dropped a report titled 'Net Zero by 2050: A roadmap for the global energy sector', which recommended an end to all new funding for coal, gas and oil development. To actually achieve net zero, it said that money had to be diverted to fund a massive expansion in renewable energy manufacture and installation. It was also explicit in saying that those working in fossil fuel production would inevitably face 'challenges'.

Joshua devoured the report in a single night, starting with the section on oil at page 101, which began:

> The trajectory of oil demand in the NZE [net zero emissions scenario] means that no exploration for new resources is required and, other than fields already approved for development, no new oil fields are necessary.

'Holy shit,' he thought to himself. 'This is huge.'

The IEA report hit the industry like artillery fire – but when he went into work, it was like there had been no impact. For about a week, it was all business as usual, until a 'town hall' meeting was called at which Santos CEO Kevin Gallagher would address the workforce. Gallagher was a short man with a tall Scottish brogue, and Joshua was 'itching' to ask him about the report – it had been all over the news, and he thought the new CEO would have to address it.

And he was right. The report was one of the first things Gallagher spoke about.

'He pretty well called it a joke,' Joshua recalls. 'He said something about how one of his colleagues had read it and it was hysterical. He said, "The world needs more gas."

'Here was the CEO, in front of the entire company, saying anyone who cared was a greenie.'

Joshua considered himself no greenie. The reality was he just liked being in nature and the natural environment. On his reading, the basic scientific reality of climate change wasn't a political issue. It was just a question of physics, as was the change it demanded. But as Gallagher laid down the law before the entire workforce, Joshua felt the CEO was talking directly to him, telling him exactly where he stood.

He left that town hall feeling flat and dejected. The 'real kicker', he says, was that a few days later he caught an interview with Gallagher on the news. During the conversation, the Santos CEO justified the company's strategy by quoting directly from the IEA report, explaining how much carbon capture and storage was needed to meet the world's climate goals.

'I remember being like, mate, you can't just pick and choose,' Joshua says. 'It was such a kick in the guts that the people at the highest levels of this organisation just didn't take this seriously.'

Joshua didn't crash out immediately; the anger simmered for a few months more, as he began to look afresh at his workplace. His colleagues were really smart people, and yet there seemed a disconnect. Human resources kept asking him and other graduates to show up to talks at universities to promote the company as a future career pathway. Their desktop screensavers included little 'BBQ facts' – factoids the company thought employees might like to share as conversation starters when standing around a burning grill on the weekend. There were 'energy solution teams' working on emissions reduction, but it started to look to him like their real function was to help sell the company, not to find solutions to climate change.

Joshua stuck around long enough for the launch of the Barossa gas project. The project had a reputation as one of the dirtiest in the country, with CO_2 concentrations across the field averaging 18 per cent, but those heading up the development had given an internal presentation titled: 'Barossa is good for climate change'.

Joshua quit the industry not long after. Others might have refused to confront the reality of climate change or their role in it, but he couldn't ignore it any longer. Instead, he packed his things and went home to the family farm. It didn't pay as well, and he might not know what he was going to do next, but he knew he didn't want to look back in ten years' time and know he had helped to build a climate-shocked future.

'It's unfortunate,' he says. 'All employees in the oil and gas industry have some blood on their hands to face up to, I guess.'

12

It Does Make Us Chumps

It was late February, and across Perth the signs for petroleum companies glowed softly over the city skyline, illuminating the mess of knotted highways and vast distances beyond. The 2022 Perth Festival's Writers Weekend was drawing to a close, and Tim Winton was about to give the final address. The four-time Miles Franklin winner and cherished West Australian author stepped up to the podium in his flip-flops, placed his papers upon the lectern, put on a pair of black, thick-rimmed glasses and began to speak.

The festival that year had been in trouble. Around the world, arts organisations from the Royal Shakespeare Company to the Edinburgh Festival had been divesting themselves of fossil fuel sponsorship. Now a group of artists wanted Western Australian institutions to do the same. Under the banner 'Fossil Free Arts WA', they had first campaigned to pressure Fringe World into divesting itself of a sponsorship arrangement with Woodside in 2021. That campaign had been a mixed success. The festival announced that it was divesting, but instead had the arrangement shifted to its parent company, Artrage. Woodside and Artrage would officially cut ties in January 2024.

Next the artists set their sights on the prestigious Perth Festival. Though the festival received considerable funding from US oil

219

giant Chevron through a decade's long partnership that started in 2005, the controversy that year centred around a performance of John Luther Adams' Pulitzer Prize–winning work, *Become Ocean*. Considered by some critics to be among the ten best classical music compositions produced in the new millennium, the performance was to be delivered by the Western Australian Youth Orchestra (WAYO). The orchestral work builds slowly, rising to a crescendo and then crashing in a parabolic arc, not unlike a wave. The idea is to capture water in motion, but Adams also intended it to serve as a warning: he says the work was inspired both by the idea that all life came from the ocean and the understanding that climate change was melting the polar ice, raising the risk of sea level rise. The implication is that all humanity may yet find itself 'quite literally become ocean'.

Naturally, it caused a stir when the program went out carrying the logo of one of Australia's largest domestic oil and gas companies, Woodside. The company was spending $16 billion to develop the Scarborough gas field off the northwest coast and wanted to build a gas-processing hub on the Burrup Peninsula. The development sparked protests as critics charged that it was clear Woodside had no intention to stop extracting gas any time soon. In fact, the company explicitly argued that the gas would help to 'firm' renewables in the energy transition.

The sponsorship deal was said to have been brokered by the West Australian Symphony Orchestra (WASO), though WAYO didn't exactly need help. Woodside had been a 'principal partner' to WAYO since 1991, with naming rights to its programs, and the orchestra regularly playing at company events, including its annual general meetings.

A group of artists and activists had repeatedly tried to arrange a meeting with WASO to point out that having a gas company sponsor an artwork about climate change was perhaps a bad idea. They were ignored. When the festival kicked off, several artists on the bill made public statements calling for the performance to be dropped. It was a big risk, speaking up. Whatever the genre, making art is a precarious living. But Winton had made a career talking about environmental issues. Where others couldn't say anything,

Western Australia's favourite literary son could speak with impunity. Winton drove right at the heart of the issue, telling his audience that the fossil fuel industry 'hasn't just occupied vast tracts of our seaway, it's colonised every level of our society'.

'And with the climate emergency upon us, the industry is not winding back. It's doubling down,' he said. 'And that, my friends, is the smouldering dumpster fire of "business as usual" in this country. If we genuinely care about preserving the conditions of life on this planet, we've got to put it out and we must do it now.'

There was no way to misconstrue where Winton stood on the issue – but he wasn't done. Taking direct aim at the fossil fuel companies' 'propaganda blitz', he described oil, gas and coal sponsorship in arts and sport as a tactic 'straight out of the Big Tobacco playbook', one that had been 'enormously successful'.

It was a claim that had particular resonance in Western Australia. Forty years before, when the state government had tried to ban tobacco advertising, it prompted a vicious response from the industry association, which claimed the ban threatened 'business freedom in Western Australia'. Their campaign was backed by the Confederation of Major Participant and Spectator Sports, a single-issue front group founded by the Tobacco Institute of Australia in 1983, which acted in concert with the tobacco industry to assert the right of sports clubs and leagues to take its money. It would count for little. The Western Australian government stood firm – though the fight wasn't without casualties. Public debate over the bill was so fierce that a fistfight erupted in state parliament and an MP was charged.

Just as sporting clubs had found it hard to give up tobacco, Winton said, the arts would struggle to give up money from 'Big Daddy gas' – but they had to. Gas companies were trying to present themselves as heroes, 'indispensable to civilisation'. They weren't courting the arts – or sport – to be nice, but as part of an 'old soft-power ploy'. In the spirit of 'collegial reflection, not condemnation', he asked his audience a simple question: why keep taking the money?

'It doesn't make us bad people, but it does make us chumps,' he said. 'All around us every day, financial institutions, super funds, shareholder groups and banks are withdrawing their patronage of the fossil fuel industry because it's seen increasingly as a bad bet, with looming stranded assets, and, in their view, it no longer passes the ethics test.

'So how is it that we in the arts community should show less creativity and moral imagination than bankers? No offence to bankers.'

It was a provocation – a direct challenge to festival organisers and the arts across the board to do better in a state dominated by the gas industry – and one that would have an effect. In October, Chevron announced it was pulling its funding from Perth Festival following all the bad press, ending an 18-year association. Arts administrators might have been stricken at the prospect of finding an alternative source of finance, but those pushing for change celebrated. Winton had been absolutely correct, they felt – though there was one point on which the celebrated author had been wrong. Australia's oil, gas and coal companies hadn't borrowed from Big Tobacco's playbook. As it turns out, the oil industry were its original authors.

The relationship between oil and tobacco was born in the parking lot of the American gas station. After petrol, cigarettes were the second-biggest selling retail item in the service stations that dotted the continental United States, and from this commercial relationship the two industries forged an alliance that continues to this day.

Early on the two orbited each other. With time, and increasingly insistent claims about cancer risk, a mutual suspicion would develop as each industry pointed the finger at the other, but at the outset the tobacco industry executives enviously eyed their counterparts in oil. From their perspective, the oil guys always seemed to have it all. They were always expanding, politically savvy and well-coordinated at the highest levels. Mostly, though, they were powerful. The oil executives had a level of influence the tobacco industry dreamt of.

The tobacco guys would get there in the end, eventually becoming oil guys themselves. By the 1970s, the chair of British American Tobacco – the world's biggest tobacco company – also served on the board of Exxon – the world's biggest oil company. Cigarette company RJ Reynolds Tobacco even once owned and operated American Independent Oil, which ran a refinery in Kuwait.

But it was the humble gas station that brought them together. Through this early commercial relationship they collaborated, hammering out cross-promotional deals to target customers at the point of sale. They even co-sponsored baseball teams – and it was this sort of association that first taught the tobacco industry the power of sport.

As a cultural institution, sport was unique. Unlike the arts, sport was accessible to a mass audience. No-one needed a university degree to learn the rules of football or watch a NASCAR race. Whatever sport a red-blooded American loved – be it baseball, or football, or drag racing – what mattered was that they felt something powerful when they watched. Every game had its own narrative and context. Sport delivered incredible highs, deep despair, moments of crisis or catastrophe and – most valuable of all – triumph. Winning – the act of overcoming an opponent – generated a euphoria more potent than nicotine. Associating a brand with this powerful positive emotion was a good way to harness the 'halo effect' – the idea that a person standing next to a saint begins to look a little saintly, too.

Beyond marketing, sport offered other, more material benefits. Speaking notes from a speech given by an unknown RJ Reynolds executive in May 1980 make it crystal clear that the tobacco industry was aware of the political power of sport:

> The political contacts we have made in sports marketing have proved invaluable. Just think about it, Reynolds sponsors over 2,400 events a year. That translates into 2,400 opportunities to put a politician in front of anywhere from 3,000 to 200,000 potential voters. An opportunity provided by Reynolds that's too good to be true for the politician.
>
> The owners of our facilities and the key players in the sanctioning bodies have also gone to bat for us when unfavorable legislation was introduced. NASCAR alone has been instrumental in killing tax increases in both Florida and Alabama.

We intend to call on these people with increasing frequency as
the environment gets tougher. We expect they will respond with
their usual enthusiasm for our company and our industry.

It was a similar story with the arts. Associating a brand with a work
of great beauty, skill or mastery was valuable for the same reasons
a company might kill to get their logo courtside. Being a patron
of the arts, however, also had other perks. If a corporate box at the
football wasn't exactly a particular politician or business associate's
thing, maybe they'd prefer tickets to the orchestra instead. Better
yet, this funding could be used to promote certain works or artists
over others. Those artists might even feel such gratitude they would
defend the industry in times of need. Arguably, a savvy-enough
operator could, with time, shape culture itself. Targeted funding
could ensure certain works, voices or careers endured as others
faded away. Throwing a little money around was a fast way to make
friends – it didn't even have to be a lot of money.

It was also a strategy that could arguably be applied to information
itself – another strategy the tobacco industry picked up from oil.
The first link between cigarette smoke and tobacco was made in
1953, when tumours grew where scientists had swabbed the back of
a lab-rat with tobacco tar. As news of the story made a splash in the
press, the tobacco executives turned to their friends in oil for help.

In 1954, a former executive with an Exxon forerunner wrote a
letter to the newly formed Tobacco Industry Scientific Advisory
board to recommend a list of scientists who had previously worked
with the oil industry on subjects ranging from smog to cancer.
These scientists would become the go-to experts for the tobacco
industry – some featuring in Naomi Oreskes and Erik Conway's
book *Merchants of Doubt* – when it sought to run interference against
proponents of the link between smoking and cancer. It was the same
model that would later be used by oil companies to run interference
against climate change.

Documents pulled out of archives by the Center for International
Environmental Law show how even as concerns about cancer-risk
grew, oil companies gave over their labs to tobacco companies.

Exxon, Shell and even the American Petroleum Institute struck deals to allow cigarette companies to tap their scientific expertise for testing of cigarette tars. Some of these labs, like the Truesdail Laboratories and the Stanford Research Institute, would also be involved in conducting the earliest atmospheric testing on the greenhouse effect.

If the oil industry taught tobacco how to fight science with science, it was the early public relations firms who acted as the lynchpin. The tobacco companies were desperate for political, social and cultural capital, and they would get it by hiring the same public relations teams the oil companies relied upon.

In their hour of need, the tobacco executives would contract what was then the oil industry's preferred public relations firm, Hill & Knowlton. The firm kept offices on the 33rd floor of the Empire State Building and would assign the same internal team responsible for looking after its biggest oil clients – companies like Texaco and Caltex – to manage the tobacco industry account. Drawing on their experiences in oil, this team would help the tobacco executives set up their own industry association, partly modelled on the American Petroleum Institute.

At their earliest meetings, Hill & Knowlton's advisers explicitly counselled the tobacco executives not to deny the science directly. Their preferred approach was simply to insist it was all just a matter of opinion, no matter how much evidence was gathered. It was the same line that the oil companies would take in confronting the emerging science confirming the greenhouse effect.

The man tasked with overseeing both the tobacco and oil accounts was Richard 'Dick' Darrow. Darrow was a former journalist who had started his career working for newspapers and radio stations in Ohio and Washington before he took a job in 1952 as vice president at Hill & Knowlton. A few years later, he would be running the firm. Darrow soon earned a reputation inside the company as an 'experienced practitioner' willing to travel the world to serve the firm's clients. One such trip would take Darrow from the lofty skyscrapers of midtown Manhattan all the way to the exotic climes of Perth, Western Australia, the most isolated big city on the planet.

During a meeting with the US president, it was impressed upon
Hill & Knowlton's corporate leadership that Caltex's operations
around the world were important to US strategic interests. The firm
was encouraged to survey the 'public relations problems in various
countries' to assist. The oil company had a joint venture with a
dinky little Australian outfit called Ampol, it was explained, and
they were supposed to be drilling at a site called Rough Range-1.
Around that time, a 'crisis' had emerged in Western Australia
involving 'some critical problems with the press'. It's possible this
referred to the controversy that began when Ampol kept news
of their oil find at Rough Range-1 secret from William Walkley
himself. When the company finally cabled major newspapers
revealing the discovery, Terry Southwell-Keely, former deputy
chief of staff for *The Sydney Morning Herald,* had called Australia's
first oil baron to congratulate him, but was surprised to hear
Walkley had no idea.

'This is the first I've heard about it,' Walkley said. 'I'll phone and
ring you back.'

Southwell-Keely would later go on to serve as Ampol's head
of public relations, but in the meantime, Darrow was dispatched
to schmooze the local press. A subscription newsletter for Hill &
Knowlton clients in the tobacco industry archives dated 23 April
1954 records how Darrow later described his work:

> A few days later, Dick Darrow returned to the office after a 40,000
> mile Australia–New Zealand trip for Caltex. His extensive travel in
> those two countries included an 850-mile overland trek with West
> Australian newspapermen from Perth to Exmouth Gulf where a
> Caltex subsidiary is drilling for oil. He blames his sunburn on that
> journey, plus 'essential' seagoing conferences with newspaper editors
> and an enforced two-day layover in the Fiji Islands for engine repairs.

In short: it was a sleaze-fest from beginning to end.

By coincidence, this same memo also records another critical
event. At the bottom of page four, the firm reported some other
good news regarding its tobacco account. The Tobacco Industry

Research Committee (TIRC), another Hill & Knowlton creation, which had offices on the floor directly below it in the Empire State Building, was making progress in recruitment.

'With the help of H&K,' the memo said, 'TIRC has succeeded in getting some of the top cancer experts in the country to join its Advisory Board.'

At the time of its formation, the ostensible purpose of the body was to 'objectively' study the relationship between cigarette smoke and cancer. In practice, its claims of independence were a smokescreen for its role as a central node in the tobacco industry's disinformation machine. Its experts would be responsible for raising doubt about the growing body of research in the name of 'good science'.

But that was not the only organisational work the firm was doing for the tobacco executives. With saccharine triviality, the memo reported that a card index file was being developed to 'keep tabs on current published views on smoking and health', a scientific bibliography of lung cancer research was being put together to help stay across the latest papers, and a fresh mailing list for contacts in industry, research and the trade press was currently in production.

This was the banality of evil Hannah Arendt would later write about. What seemed ordinary and prosaic would prove to be the basic foundations of the tobacco industry's desperate effort to stave off litigation. Decades on, it is impossible to put an accurate body count to these actions.

And between these worlds stood Richard 'Dick' Darrow. He was like a common ancestor in an evolutionary tree linking together the US oil industry, the tobacco industry and the Australian oil industry. The man who taught tobacco companies to lie like the US oil industry — to lie by omission, to palter — was also midwife at the birth of the Australian oil industry. He had partied with reporters and sailed with their editors. Walkley later complained bitterly that Darrow's efforts had meant Caltex received credit for the find at Rough Range-1 exploration well to the detriment of Ampol, and would campaign for Darrow's replacement, but that was no setback.

In fact, Darrow seems to have found his time in Australia so agreeable that in July 1954, a few months after he made his return, Hill & Knowlton opened an Australian office in Sydney.

Darrow may have been a successful man, but he would not succeed in growing old. In 1976, at age 60, he died from cancer.

William Walkley was nothing if not a generous man. Dressed, as usual, in his trademark powder-blue suit and matching tie, he pulled his chair around to sit next to his guest, almost shoulder to shoulder when he handed the artist a fistful of cigars. The year was 1956, and they were dining at Romano's, Sydney's ritziest restaurant, a regular haunt for the city's movers and shakers – the kind of joint where the wait staff wore full tux and bowtie.

It was after dinner, and Albert Namatjira clutched the cigars in his hand, one for each person at the table. The Arrernte man, one of Australia's most famous artists, cut a charismatic figure. His success increasingly allowed him to slip the restraints of a segregated Australia and brought him to places from which he might otherwise be barred, and he had no trouble holding his own among the Sydney elite. The artist shared a joke with the country's first oil baron, and Walkley responded with a big, gregarious grin. Taking a cigar in hand, the oil baron held it out as if to explain its quality and construction.

At the table with them was Namatjira's 18–year-old son, Keith, best-selling Australian writer Frank Clune and Bert Gardiner, a taxi driver from Alice Springs whose job would be to drive them home. Clune had arranged Namatjira's trip south, which required getting permission from the Northern Territory Administration's Welfare Branch for the artist to travel. Their purpose was to collect a new truck – a brand new 1956 Dodge Kew Coupe Ute. Back home, the artist hauled around his art supplies in an old truck with a wooden tray and the words 'Albert Namatjira, Artist, Hermannsburg' printed on the driver's side door. It was like a signature.

When it was clapped out, Namatjira struck a deal with Walkley, who was a collector of his paintings, to replace it. Parked outside, the new truck was a fine machine. A fresh inscription had even been painted on the driver's side door – the words 'Albert Namatjira, Artist, Alice Springs'. But this one had an addition. Down by the rear driver's side wheel well were the words: 'This vehicle presented by Ampol.'

In some ways it was no different to a basketball player getting a Nike sponsorship, or a football player lending his name to a used-car dealership. Namatjira was the best, and his truck iconic. Under the deal, everyone seemed to win. The replacement truck came fully loaded with art supplies and gifts from the company. With this gesture, Ampol ensured that whenever the artist posed for a photograph against the truck, the entire country would remember William Walkley's generosity.

The story these photos tell goes beyond art history; what they demonstrate is the extent to which Australia's oil industry understood the power of culture and celebrity from the very beginning – and how they actively sought to cultivate it. With Ampol, Walkley would go on to set the mould for successor companies and executives, says Belinda Noble, a former journalist who now monitors public relations firms and their work with fossil fuel producers through CommsDeclare.

'They sponsored everything that moved,' she says.

This is no exaggeration. When it came to art, Walkley collected Namatjira's paintings and his company sponsored Musica Viva, which supports chamber music in Australia. He also dabbled in media. Walkley would also give his name to the highest prize for journalistic achievement, the Walkley. Founded in 1956 at the suggestion of Southwell-Keely, Walkley would present these awards personally until he was stopped by ill health and it was taken over by the Australian Journalists Union. Today it is considered Australia's equivalent to the US Pulitzer. On one analysis, only a single award has ever been given to a story directly addressing climate change.

In the realm of entertainment, the company sponsored a show on 2GB radio, hosted by New Zealand-born singer Jack Davey. This arrangement brought the *Ampol Show* into every living room in Sydney, featuring quizzes, mystery melodies, tongue twisters and limericks.

But it was sport where Walkley was most active. Motorsport was an obvious target for the man who owned the nation's petrol pumps, and Jack Davey himself would race in the Ampol round-Australia reliability trials in 1956, 1957 and 1958. The company also struck sponsorship arrangements with fishing, polocrosse, rally driving, rugby, football, soccer and surf lifesaving clubs. Ampol would give its name to a soccer tournament (the Ampol Cup), a golf tournament (the Ampol Tournament), and a soccer trophy awarded to the winner of matches between Australian and Chinese soccer teams in the mid-1980s (the China–Australia Ampol Cup). The focus on soccer was in many ways cynical: Walkley and Southwell-Keely were aware a new generation of European migrants were making their way to Australia and were a rapidly growing part of the workforce. Sponsoring the sport helped win their loyalty, which may explain how Walkley ended up elected to head the Australian Soccer Federation and the Oceania Football Confederation, where he negotiated Australia's re-entry into FIFA.

None of this was altruism, Noble says. It was just good business.

'Influence is the act of bringing other people around to your point of view,' she says. 'You can do it on a personal level, but you can also do it on a mass level. When you do media and sponsorships, you're attempting to bring people around to your point of view at a mass level.

'Petrol, tobacco, public relations – all these things grew up together after the Second World War in a world with a new aspirational and affluent society.

'We grew up with these messages. That's why they're so hard to nudge now. People have a warm, fuzzy feeling about Ampol because of the petrol stations and the advertising.'

All of it was a smokescreen for Walkley's politics. He was everywhere, and that meant he wasn't just *a* top bloke, he was *the* top bloke, and it would have taken genuine bravery for anyone to publicly disagree. Throughout his life, there would be no real critical evaluation of how he thought and acted, though it is possible to get a sense of how he thought from an article he was invited to contribute to *The Sydney Morning Herald* in April 1961.

As part of a series, Australia's first oil baron was asked the question, 'What would you do if you were the ruler of Australia?' The White Australia Policy was being dismantled at the time, and Walkley answered the *Herald*'s question by calling for the wall to go back up.

'Australia is under-populated and under-developed,' he began, 'This is the crux of our political problem. Unless Australia's deficiencies in manpower and economic resources are made good in the shortest time possible, it will cease to be a white man's country.'

As populations in Asia increase, Walkley said, Australia will face process to 'open our gates to coloured migrants'. He might have been friendly with Albert Namatjira, but the removal of racist controls on immigration appeared to unnerve him.

'Today Australians are but a drop of white in a sea of colour that teems with more than 1,200 million land-hungry Asiatics,' he wrote.

It was a statement that would go overlooked until 2023, when journalist Osman Faruqi dug it out of an archive. The time it took for anyone to cast a critical eye over his views spoke to the depth of his influence across the country, a legacy that was built thanks in part to Ampol's sponsorship program.

It is unknown whether Walkley and Southwell-Keely relied on sage advice from Hill & Knowlton. Ampol merged with Caltex in 1997, and historical documents either no longer exist or are unavailable. What is known is that Hill & Knowlton was actively courting the Australian oil industry at this time. APEA considered hiring the firm in 1961, but was ultimately too cheap to go through with it. The £5000 retainer quoted was considered 'too dear', and the association countered by asking what Hill & Knowlton 'could do for £500'. Still, the association resolved to hire Hill & Knowlton from 'time to time'.

Fast-forward half a century since Walkley showed the industry how it was done, and these kinds of sponsorship deals are everywhere. Wherever an oil or gas company wants to soothe or employ people, it has sought to buy goodwill. According to a list compiled by author Penny Tangey, there were over 500 known industry partnerships across Australia at the end of 2022, with oil, gas and coal companies having struck arrangements with sports, arts, environmental, community and educational organisations across the country.

It's impossible to know how many of these sponsorship arrangements exist at any given time. New deals are being inked every day as old ones fade – and most are shrouded in secrecy. The terms of these arrangements are rarely made public, with many thought to contain non-disclosure and non-disparagement clauses that handcuff organisations or their members from criticising their sponsors. This lack of transparency also makes it hard to assign a dollar value to these deals. One review by researchers working in partnership with the Australian Conservation Foundation estimated the value of fossil fuel sponsorship in Australian sport alone at between $14 million and $18 million a year.

The result is a convoluted and opaque web of funding arrangements that crisscross the nation, binding institutions to the fossil fuel producers on whom they have come to depend financially. In some remote parts of the country, where government funding has been slowly withdrawn by penny-pinching authorities, fossil fuel producers have stepped into the resulting vacuum. The result is almost a modern iteration of the company town.

By any objective measure, the tactic has been a stunning success, one used again and again for nearly half a century without opposition – until that is, artists themselves began to revolt.

Standing in the café bathroom, Joana Partyka took a slug of mezcal from a flask and a moment to steady herself. Above her, the fluorescent lighting buzzed as the 38-year-old climate activist looked at herself in the mirror. She wore a long, roomy black coat with deep pockets to cover a white t-shirt with the phrase 'Disrupt Burrup Hub' printed on the front. Earlier that morning, before she left home, she had slipped a can of yellow spray paint into her coat pocket.

The coat made her uneasy. This was Perth in mid-January, not exactly coat weather. On her way over, she had been careful not to call attention to herself. Had she lingered, anyone who noticed her on the street and considered it for longer than a minute would have thought her get-up strange.

Partyka reapplied the red lipstick she always wore to make herself feel powerful in tense moments, looked herself in the eye and whispered that it was all going to be okay – that the act of vandalism she was about to commit was important. Fluffing her hair, she walked back out into the café, where her support crew were going over the plan one more time. When they were done, she picked up the large sketchbook that held the stencil, walked over to the Art Gallery of Western Australia and straight into the national spotlight.

Partyka can't quite remember whose idea it was to brand the beloved 1889 Frederick McCubbin painting *Down on His Luck* with the Woodside logo, or how exactly she ended up actually carrying it out. In October 2022, activists with Just Stop Oil over in the UK had thrown tomato soup over the perspex covering of Vincent van Gogh's *Sunflowers* at the National Gallery in London as part of their campaign against fossil fuel expansion. The action had inspired similar protests across the world. The tomato soup was the kind of detail that appealed to nerds who knew enough art history to recognise the allusion to Andy Warhol, but it had never sat right with Partyka. As a political statement, it seemed inaccessible and vague. It didn't call attention to anything. Worse, all that goopy, drippy non-Newtonian fluid might actually have damaged something.

Later on, in the aftermath of the Perth protest, it would be reported as if Partyka and her crew had damaged the painting itself. This was false, though the activists weren't exactly in a rush to correct lazy reporters, as it gave the story legs. Truth was, the damage was an illusion. They had carefully engineered the whole spectacle specifically to avoid damaging the painting, which remained safe behind its security panel.

In the weeks leading up to the action, Partyka and a group of several others had been kicking around a few ideas. Drawing inspiration from Just Stop Oil, they wanted to call attention to industrial expansion on the Burrup Peninsula in the far north of the state. The action was organised by a group calling itself Disrupt

Burrup Hub, and Partyka, a national communications officer with the Greens, had volunteered to take part when a call-out went around. Until that point, she had mostly been involved with Extinction Rebellion, having picked up her first arrest in an act of civil disobedience in October 2019. She worked in politics, but she had joined the protest movement when she grew disillusioned with the necessary compromises of parliament that slowed the pace of change. Though it felt like the protests had gained some momentum, it had stalled with the pandemic. With the interstate lockdowns having ended and the borders open, she was looking to restart, and the Burrup Peninsula seemed a worthy cause.

They already knew which painting they wanted to target. The McCubbin was a colonial-era masterpiece, valued at $3 million in 2011, which ranked among the gallery's most prized works. Around that time, it was on display in an exhibition about the dispossession of Indigenous people. To a European eye, it depicted a tired old prospector at rest in a natural landscape; to First Nations peoples, it showed a man trying to extract mineral wealth from land that wasn't his.

During the early planning stages, they kicked around ideas about how they could approach it. As both an artist herself and the person carrying out the action, Partyka's main condition was, that whatever they did, the painting couldn't be damaged. With soup out of the question, someone pitched the idea of spray-painting the Woodside logo on the painting's protective shield. Instantly, something clicked. There was a simple elegance to the notion. Woodside was going around trying to slap its logo on everything anyway, so why not give them a hand? The message would be obvious, and it also tapped into an established artistic tradition. The Situationists, a band of idealistic French radicals, had similarly blended art and politics together with their sloganeering during the Paris riots in '68, and in a separate example, people still regularly turned up to piss in Duchamp's toilet.

Unlike the soup protests, it helped that the symbolism lined up perfectly. It was possible to draw a straight line between cause, intent and spectacle. Extinction Rebellion internally referred to those who engaged in acts of protest as the 'actor' or 'artist'. Meanwhile, the rock art on the Burrup Peninsula being threatened by Woodside's Scarborough development was equivalent to a 60,000-year-old open-air art gallery. Its petroglyphs are so much older than Rome, more ancient than the Sphinx, yet Woodside wanted to build a giant processing hub right in the middle. The cumulative waste gases from the plant and neighbouring operations would vent into the air and then settle, changing the chemical composition of the rock face until the petroglyphs began to corrode. Partyka was just emulating Woodside, who claimed their chemists had investigated the issue and found there would be no effect on the rock art. Unlike Woodside, Partyka would be spray-painting the logo over the protective coating of the painting, not the actual artwork. No harm, no foul.

It was a good idea, better than soup. That was enough. From the moment it was suggested, she blocked out any thought that didn't relate to the specific tasks needed to actually carry it through. Until it was over, she would have no real cognitive capacity for anything else.

The activists drilled the protest until it was second nature. When the day came, they all knew their roles. Stepping through the doors to the art gallery, Partyka didn't go straight to the painting. That would be too obvious and leave her open to interception. To hide in plain sight, the activists staggered their entry. They had built in time to meander between the works before converging on one central point. Floating through the galleries, Partyka stopped to sketch different pieces as she gradually worked her way towards their target. By coincidence, a small lounge was positioned right in front of the McCubbin, so she sat down and began to sketch while she waited for the rest of the team to get into place.

Taking off her coat – carefully, so as not to call attention to the rattle of the spray can in the pocket – she eyed off the gallery attendants and security guards as they made their rounds. She continued to sketch as she looked for an opening but soon grew frustrated. Every time she thought she had an opportunity, another person would appear.

When her window finally came, she found herself moving like an automaton. Pulling the stencil from her sketchpad, she walked over to the painting, blasted it with yellow paint and superglued herself to the wall. As she did so, Ballardong Noongar man Desmond Blurton, Indigenous spokesperson for Disrupt Burrup Hub, livestreamed a speech demanding that the Woodside project be stopped and Indigenous heritage be respected.

'As I stand here today, artwork that is sacred to my people is being destroyed in Western Australia,' Blurton said. 'Woodside Petroleum is the largest fossil fuel project in Australia; they are destroying the ancient Murujuga rock art.

'We demand no industry on the Burrup now.'

When he was done, it was Partyka's turn to speak.

'Woodside is happy to slap its logo on everything, including some of our most significant cultural institutions in this country, particularly in WA, while they spray toxic emissions on sacred rock art, which is also irreplaceable,' she said.

'We must stop more industry on the Burrup, and we must disrupt Burrup Hub now, before it is too late, before we lose this irreplaceable rock art.'

LATE PERIOD

13

We Are Here to Listen

The wind made it hard to breathe. It was only a 25-metre climb up the steel steps of the fire tower, but the front sweeping in from the west hit hard. The feeling was not unlike suffocating. As we stood at the top, looking out over the Pilliga Scrub in central New South Wales in early September 2022, we had to lean into the wind so as not to feel like we would be blown away.

According to the Bureau of Meteorology, a powerful cold front had moved across South Australia the night before and was starting to make its way into New South Wales. When it landed, it would bring severe thunderstorms and up to 40 millimetres of rain in some parts, creating a risk of flooding. It was only noon, and we had yet to visit the actual drill site, but the sky had already begun to bruise. Everyone knew we were on a clock.

The view from the top helped put a few things in perspective. A panoramic compass circled the ceiling, using individual marks to show the direction of nearby towns – Narrabri, Bugaldie, Coonabarabran. At 5000km², this was the largest remaining scrub forest west of the Dividing Range. The size of the place was only matched by the size of the investment Santos was making here: under plans for a $3.5-billon Narrabri gas development, the company planned to sink more than 850 new gas wells.

By 2022, the Narrabri project had been a decade in the making. Santos sold it to the public on the promise it would bring security and wealth. An alleged gas shortage had developed up and down the east coast during 2022, and the company had promised to keep Narrabri's gas exclusively for domestic Australian use.

To pull it from the ground, Santos planned to scour clean a total of 10km^2 of state forest and privately owned farmland. Drilling would dig 1200 metres below the surface to extract 1500 petajoules of gas. When complete, the wells would pockmark the Pilliga.

It was a pitch that Narrabri's then mayor, Ron Campbell, found persuasive. It was reported in August 2022 that he was in discussions with Santos for his waste disposal company to handle 33,600 tonnes of salt and chemical waste the company's drilling operations would produce each year. This was the equivalent of four Eiffel Towers a year for the lifetime of the project. Campbell didn't think he might be compromised by the commercial relationship. When he was asked about it, he said the state government was responsible for giving Santos the go-ahead, not the local council, and that he likely wouldn't be mayor by the time work started. And he was right; in September 2023, he was deposed as mayor but remained a councillor.

It just so happened all this money making would take place on the traditional lands of the Gomeroi people, who were actively fighting the development. At the time I visited, some 23,000 submissions had been lodged against the proposal during public consultation. It was a record, but the then federal Coalition government had waved away the public opposition. After the pandemic, it had a plan to restart the economy in a 'gas-fired recovery' and it was not going to allow community opposition to interfere. In October 2021, Santos was given the green light, a decision the company 'welcomed', saying it looked forward to 'getting on with our work in regional New South Wales to create jobs, drive investment and bring long-term energy security to the state at competitive prices'.

Santos already had demonstration wells operating in the area, and we had come to see them. These facilities were tucked away in the scrub, and the Climate Council, an advocacy group, had arranged a junket to visit the facility. Aware of this, I independently organised for myself and a photographer to be in the area so we could tag along.

The morning had begun as planned. Having linked up with the contingent from the Climate Council, we drove in convoy along a sealed road somewhere behind Coonabarabran towards a drill site when the lead car pulled off into the scrub. A short way down a dirt road, the lead car came to a halt and out climbed Suellyn Tighe. Tighe is a Gomeroi woman from a respected family who was participating in the Indigenous nation's opposition to the project. Standing before the assembled group, she gestured towards a resplendent field of yellow wattle.

Moving among the blooming native trees, she pointed out features of the landscape, explaining how everything was interrelated. The wattle was a calendar plant, she said. When it bloomed, it signalled that the water in nearby rivers was warm enough for the first yabbies to appear and that native fruits were beginning to ripen.

This was how she knew climate change was real and happening. She could read it in the landscape. Things were either growing too fast or too slow, and the usual rhythm of the plants and animals' interactions was breaking down. The native fruits were arriving early and were nearly mature by the time the wattle flowered. Up in the trees, the birds were being tricked into nesting earlier. Some of the smaller species, like finches and wrens, had taken to raising young twice in a year. Back when the droughts came, Tighe said, their numbers had been decimated. The thirsty trees' thinning canopy offered little shade from the heat or protection from predators. Chicks died in the blistering summer sun or were picked off by eagles and hawks, whose own numbers were multiplying.

Traditionally, Tighe explained, the Gomeroi divided the year into eight or more seasons, depending on the specific region they were in. The seasons had always come and gone like clockwork, telling them when to move, when different species would arrive or grow bountiful, and when to burn the landscape to 'clean' it. Everything depended on the nature of the seasons, but now the clear delineation between one season and the next was growing fuzzy, more incoherent. To Tighe, this was all so obvious that it was hard to understand how others didn't see it.

'I'm amazed that people aren't looking out the window and seeing what's happening,' she said. 'They're looking through a cataract lens.'

For Tighe, climate change intersected with the threat posed to their sovereignty, self-determination and Country by the incursion of the oil company. As humanity was causing climate change, and climate change was interfering with Country, it also interfered with their ability to care for it and, by extension the Gomeroi.

As part of their decade-long fight against the project, the Gomeroi had set up a citizen science project to monitor the industrial operation and started running site tours to the well sites, taking curious reporters, activists and researchers on fact-finding tours to walk the perimeter. Santos also ran its own trips, organised from a shopfront on the main strip in Narrabri. The nine-to-five operation served as an information point for curious locals and a convenient way to show it was consulting with the community. It was brightly lit, with freshly painted white walls, and staffed by the kind of people who wear puffer jacket vests, clean flannel shirts and RM Williams boots. The company even set up a dedicated website through which to offer its own site tours on the third Thursday of each month. Its slug line read: 'Your land, your choice'.

Santos had initially approached the Gomeroi promising work and riches. Then, in late 2011, the Gomeroi filed a native title claim over a sprawling area of New South Wales that included all of the Pilliga, meaning that Santos was required to consult with the Gomeroi before forging ahead on the project. Fast forward a decade, and the company's patience ran out. In March 2022, during the pandemic, Santos applied to the courts to allow them to proceed on the project without Gomeroi consent. In response to the filing, the Gomeroi went to a nation vote – a vote of the whole community – to decide whether to accept an offer the company made to secure permission to drill. The result was a resounding 'no'. Of the 162 Gomeroi participants, only two voted in favour of letting Santos proceed, and four abstained. Everyone else voted against it.

If there was any lingering doubt about the true feeling among the Gomeroi, they made their feelings perfectly clear two months later at the company's 2022 annual general meeting. About halfway through, a distant cousin of Suellyn, Dorothy Tighe, stood up from the audience and approached to serve the company's chair, Keith Spence, with a cease-and-desist notice. Though it had no legal effect, her symbolic action brought the confrontation between the community and the company into sharp focus.

But having invested ten years and hundreds of millions of dollars in the Narrabri gas project, there was no way Santos was going to let the decision of the Gomeroi stand. And, at first, Santos won. The Gomeroi naturally appealed and in March 2024 they would win this appeal but when we visited in September they were still waiting on the outcome. It was a moment that exposed the pretence of collaboration between the company and First Nations people.

Gomeroi man Raymond Weatherall described it as history repeating. 'They're coming to take our country again, just this time without the muskets,' he said.

After several stops to get acquainted with the landscape, the convoy made its way in the direction of the Pilliga and Santos' drill site. Actually getting there required driving down a series of long dirt roads that crisscrossed the scrub. Another Gomeroi woman, Deborah Briggs, led the tour around the perimeter of the facility. She had grown up in the Pilliga, where her father had worked as a sleeper cutter for 50 years. Now she helped run the citizen science project monitoring industrial activity at the site. Signs at the locked gate warned us as we approached that we were being watched, through closed-circuit television and on trail cams Santos had mounted in the area to keep watch for activists.

Looking through the chain-link fence, the well site itself was nothing special. The brush had been rolled back, the ground levelled and covered in gravel or asphalt to make a work surface. Tucked away off a dirt road, the site felt lonely. For all the talk from Santos about how the expansion would create 912 full-time jobs, there wasn't a soul around. Everything was monitored remotely by small, off-site teams who checked the cameras and took the readings from the machines before responding to any situation as it develops.

As we walked around the perimeter of the facility, we passed an area surrounded by rent-a-fence. It jutted off from the main facility, setting up a cordon to keep people from entering an area contaminated by a spill a decade prior when it was under different management. The area had been contaminated when a leak from the treatment plant caused 10,000 litres of wastewater to overtop a pond and flow freely into the bush. An effort had been made to rehabilitate the area, but the vegetation was thin and sparse compared to the surroundings. Further on along the fence, the scrub grew thick and lush. To European eyes, this might be a selling point, but without the active back-burning that used to be conducted by First Nations people in the area, the landscape was overgrown and laden with fuel, a potential bushfire risk. Towards the rear of the facility, the concrete walls of an evaporation pit rose. A huge storage tank, the colour of khaki and surrounded by a shallow, plastic-lined moat, held the gas extracted from the well. The well itself was an unremarkable cluster of pipes off to one side that plunged deep into the earth and the coal seam below.

With her back to the chain-link fence, Briggs addressed the small group of reporters. A sign over her shoulder read: 'PRIVATE PROPERTY: DO NOT ENTER'. She spoke about how the company's plan to build 800 similar sites threatened the land and water. She was afraid of another spill, or a bushfire tearing through the area and igniting the gas being pumped from underground.

It was around then the rain began to fall. It came softly at first but began to pick up as we worked our way back to the car. As the convoy left the site, my photographer and I broke off and drove the long way to our motel. The rain beat a familiar rhythm on the windshield. Two years ago, not far from here, the mighty Namoi River had run dry, and the catastrophic Black Summer bushfires tore their way south from the Queensland border down into Victoria. Then the rivers of New South Wales had grown swollen as torrential rains fell on dry, parched earth and began to flood. The public had responded to the escalating climate crisis with righteous anger. There was an expectation that something had to change. In other parts of New South Wales, new gas developments had been stopped by opposition movements. Now it was the Gomeroi's turn.

It had been forty years since the Yungngora had squared off against the oil company at Noonkanbah Station, but the legacy of that encounter reached down through the decades all the way to Narrabri. The guidelines APEA had originally provided to state and federal governments in the early 1980s had been rewritten, updated and republished in 1988 and 1990. This era of self-regulation ended in 1992, when the *Mabo* decision upturned 'terra nullius', the foundational lie of the Australian legal system. When the *Native Title Act 1993* was written in its wake, APEA took a backseat but was still on hand to offer helpful amendments.

When it came to resource development, the processes the Act laid out were supposed to manage the interaction between resource companies and Indigenous communities, but the reality was that those processes were weighted towards a specific outcome.

It came as no shock to the Gomeroi when, in December 2022, the National Native Title Tribunal ruled in favour of Santos. In its decision, the tribunal found the Gomeroi had failed to prove the Narrabri development would have 'grave and irreversible consequences for the Gomeroi people's culture, lands and waters and would contribute to climate change' in a way that would infringe on their rights as set out in the *Native Title* Act. In a 358-page decision, tribunal president and former Federal Court judge John Dowsett found Santos had negotiated in good faith and did not agree with the Gomeroi's claim the project would worsen climate change.

The core of this decision involved a simple reality: the law didn't give him the authority to consider the impact of climate change, or the means with which to do so. As written, the law generally set up a process of consultation with Indigenous communities, requiring certain process and procedures. So long as a company engaged in good faith with this process, they satisfied the requirements.

This was also true when it came to potential harms from

climate change. So long as planning authorities had given it some consideration, that was enough. What it came down to, essentially, is whether the constant stream of cash that would be generated by drilling 850 gas wells would be outweighed by the potential catastrophic harms of climate change.

To help their case, the Gomeroi enlisted the help of one of the country's most respected working climate scientists, the late Australian National University Professor Will Steffen. Over his storied career, Steffen had advised government departments, served as a member of the Australian Climate Commission and contributed to research on irreversible climate 'tipping points' that threatened devastating consequences if reached.

Unfortunately, Steffen appeared to have been poorly briefed about the project. In giving his evidence, Steffen thought the project involved fracking when it did not. Having demonstrated this misunderstanding, Santos' lawyers successfully used the stumble to attack his credibility as an expert witness. In giving his decision, Dowsett dismissed Steffen's concerns about climate change and the need to end fossil fuel production, as the opinion of 'one scientist' did not outweigh the efforts of the Independent Planning Commission which 'cannot simply be dismissed':

> Professor Steffen states his views that Australia could only meet the goals of the Paris Climate Agreement if it bans all new fossil fuel developments. That view may, or may not be correct, but the question is not primarily for scientists. It is for our political leaders to decide the extent to which we should seek to contribute to such goals.

During argument the Gomeroi asked that any calculation of the economic benefit of the project take into account the cost of damage resulting from climate change. Santos responded by arguing that 'it would not be possible to ascertain the proportion of any damage attributable to climate change' and that there was no evidence concerning its impact on the 'preservation and continuity of Gomeroi culture'.

In the end, Dowsett agreed with Santos.

The crux of the issue, according to Dowsett, had something to do with measurement. Because climate change was a 'worldwide phenomenon' that was 'not directly attributable to the extent of greenhouse gas emissions in north-western New South Wales', Dowsett held it was not possible to trace the harms it caused. In contrast, the benefit of an individual gas well was 'capable of predictive calculation' which meant 'it is easier to calculate the benefits of the project than to calculate the extent of damage as the result of its greenhouse emissions.'

As Dowsett himself explained:

> Likely benefits include employment for workers engaged for the Narrabri Gas Project, indirect employment for suppliers and contractors, royalties and tax revenue for the State, tax revenue for the Australian Government, and the ongoing availability of gas for households and businesses in New South Wales and Australia.

In plain English, cleansing the landscape of trees or blowing up an ancient rock formation was a measurable environmental harm, but it wasn't obvious to Dowsett what proportion of the damage caused by climate change was attributable to a gas well out the back of Narrabri. If it was not possible to get out a metaphorical ruler and measure this damage, it was also not possible to predict the harm.

And as Santos' lawyers argued, if there was no risk, there was only reward.

Dowsett appeared to leave open the possibility that future reviews could examine the impact of projects on climate change, noting that 'there are conflicting views concerning climate change and knowledge is rapidly expanding' – though he did not explain what those 'conflicting views' were. In the meantime, he pointed to the conditions placed upon the Narrabri project by decision makers as an illustration that the Department of Planning, Industry and Environment had given the issue some thought. At the same time, he took on face value arguments from the department that gas was a cleaner burning fuel than coal – a common industry talking point dating back to the 1970s that did not acknowledge how gas, as methane, was a potent greenhouse gas in its own right:

> The Department reasoned that the Narrabri Gas Project will provide an opportunity to meet domestic energy needs and, where it displaces coal-fired electricity generation, the coal seam gas power will produce a net reduction of approximately 50% in greenhouse gas emissions. However it is noted that such saving would be lost if the gas were to be exported as liquefied natural gas. The Department recommended a condition of domestic supply only, and other conditions related to minimizing greenhouse gas emissions from the project.

Dowsett's reasoning would be rejected by the Full Federal Court in March 2024. Until then, the decision was considered a clear win for Santos and a blow to the Gomeroi – though it had not been an unexpected outcome as there was always a sense of inevitability to dealings with the tribunal. Dr Lily O'Neill, a lawyer and senior research fellow with Melbourne Climate Futures, says the result is more evidence of how, in this particular area of the law, the scales have been weighted against Indigenous communities. Out of some 30,000 or so disputes heard on applications of this kind, she says only three decisions have come down in favour of Indigenous people.

'Essentially the *Native Title Act* was written to allow these projects to go ahead,' she says. 'The fact that Indigenous people, when they go to the tribunal, can't stop these companies from doing what they want is not surprising, because that's how it was written.

'When you're a traditional owner who goes to the tribunal, you know you're going to lose, because only three other groups have ever won.'

Around the same time Santos was chalking up a win in Narrabri, the company swallowed a stinging loss in the Timor Sea. It was a legal fight that had started in 2020, when Santos took over the Barossa gas field from ConocoPhillips just off the Tiwi Islands. The field lay offshore, 100 kilometres off the Tiwi coast, a remote part of the country that was difficult to reach and which lacked any existing infrastructure. ConocoPhillips had been pottering about the area since 2004, but the more they looked, the more they had come to an uncomfortable realisation. There was gas there all right, but it had a relatively high level of contaminant CO_2 – around 18 per cent across the field. Once this became known, Barossa earned a reputation as one of the dirtiest gas developments in the country.

If the leadership at Conoco had any misgivings about the project, they were not shared by Santos, where upper management framed the company's willingness to take on the remote development as a marker of its engineering prowess. All that contaminant CO_2 still represented a problem, but the company had a plan to deal with it: capture the CO_2 output and bury it in the depleted Bayu–Udan gas field in Timor-Leste waters – an expensive procedure that relied on technology that has so far not been shown to work at scale in Australia. Afterwards, the gas would be run through a 350-kilometre undersea pipeline to Darwin for processing.

This pipeline just so happened to run right past the Tiwi Islands, home to 3000 people, and closest to the island home of the Munupi clan group. The Munupi owned the northern beaches closest to the drill site and used the area for fishing and hunting. At law, Indigenous communities had no legal right to sea country from a point three nautical miles beyond the coastline of their traditional country. When Santos checked its law books, the regulations said they only had to consult with 'relevant affected persons' before proceeding. The company interpreted this to mean people with 'legal' rights to the area: commercial fishers with permits, land owners, governments, councils – but not native title holders like the Tiwi Islanders, and certainly not the Munupi. During consultation, the company spoke to several groups, but the extent of consultation with the Tiwi Islanders amounted to a notice sent to an unmonitored email account.

Job done, the company laid its money down in 2021. This proved a critical mistake. When the regulator gave the company approval to go ahead, a senior Munupi man, Dennis 'Murphy' Tipakalippa, filed a Federal Court appeal against the project with Environment Centre NT as his legal representatives. In a statement made at the time, Tipakalippa accused the company of being too afraid to face the Munupi.

'They never came to me in person or face to face,' he said. 'I think they couldn't face my people.'

To the surprise of everyone, particularly Santos, the Munupi won. The company had thought they only had to speak to those with legal rights, but the judge disagreed. When the decision came down, Justice Mordy Bromberg found the Munupi were, in fact, 'relevant affected persons' and should have been consulted.

The result was a disaster, especially for Santos CEO Kevin Gallagher. Under the terms of his contract with the company, Gallagher was supposed to receive a $6 million bonus at the end of 2025, were he able to get Barossa and other projects off the ground. But the Federal Court ruling, which initially appeared a relatively simple decision, proved to have wider implications. Suddenly there was a question about whether the entire Australian petroleum sector was correctly fulfilling its obligations to consult on new projects. In response, the National Offshore Petroleum Safety and Environmental Management Authority (NOPSEMA) began reviewing other developments to ensure they complied with regulations, as did individual companies. This one decision was rumoured to have cost the broader Australian oil industry tens of millions of dollars a day as regulators and companies doubled back to check their paperwork.

And things only grew worse for Santos. The company had been so confident of victory that it had not even bothered to wait for an outcome before it started construction work. Even as the court case was playing out, the company began drilling eight test wells in the area. In a statement following the decision, the company announced that work on the Barossa Gas Project was 'approximately 46 per cent complete' and described the decision as a 'disappointing outcome'. Santos responded to this setback by issuing a statement calling for the government to 'urgently' rewrite the rules to address 'approval uncertainty'.

'Project approval uncertainty is a public policy issue that should be urgently addressed by Australian governments to reduce risk for trade and investment in projects around the country,' it said.

Faced with the prospect of starting over, Santos appealed, but Bromberg's decision was upheld in December 2022. That was not all: a month later, in late January 2023, the company was quietly planning to start building the undersea pipeline out to its field but didn't bother telling the regulator. A week before work began, the offshore regulator, NOPSEMA, raided Santos' Adelaide offices for evidence confirming the company's plans. The regulator then blocked the company from starting construction and ordered it to check for culturally significant sites on the seabed.

None of this meant the end of the project. Santos would overcome objections over its pipeline work in January 2024 when the courts delivered a resounding decision in its favour. On the drilling work itself, the Tiwi Islanders had consistently expressed their opposition to the project, but for Santos that was no deal breaker. With its money on the table, Santos was already committed and, under the current law, all they had to do was to be seen consulting as widely as possible, just the as they had done in Narrabri. There was no obligation for them to act on what the community was telling them, or to withdraw from the development in the face of opposition.

And so, by June 2023, Santos had its teams out in force across the Tiwi Islands, holding 'listening sessions' to hear out community concerns. At these sessions, as if to be sure there could be no mistake, the company put up banners that bore the blue Santos logo and carried the slogan: 'We are here to listen.'

14

Here They Come

A smattering of grey dashed the blue April sky the morning Violet CoCo and three other Fireproof Australia activists prepared to make their final approach on the Sydney Harbour Bridge.

In the seven weeks leading up this moment in early 2022, a series of small protests had been taking place at different times, on different days, each organised by different groups. People sat down on roads in front of fossil fuel export terminals, blocked exit tunnels on a highway and invaded the pitch at football games. Each struck with precision and speed, turning the city's infrastructure against authorities and attracting the attention of the media. The activists called this skirmish 'dancing'; their aim was to be unpredictable and everywhere, while also maintaining consistent pressure. As the authorities grew increasingly incensed, the media would naturally cover the latest disruption with corresponding interest. Even hostile reporters who clutched pearls over the hold-up to morning traffic would find themselves talking about climate change when explaining to their audience why these hooligans had engaged in such brazen lawlessness.

The Harbour Bridge action was conceived as a defiant response to the state government's heavy-handed reliance on police to repress the climate cation campaign and the harsh sentences being handed out to activists in the courts – including CoCo's then partner.

Next to CoCo in the passenger seat of a rented truck sat Alan Glover. Glover was an imposing figure, 191 centimetres tall and square shouldered. An actor and comedian, he also had 44 years' experience as a volunteer firefighter. The Black Summer bushfires, which were bigger and more monstrous than anything he had encountered before, had scarred. It was this experience that motivated him to join the action on the bridge.

A tense silence dominated the cab as they drove up. Glover was nervous. He had never been arrested. The New South Wales state government had set up Strikeforce Guard to respond to the ongoing protests, and both CoCo and Glover assumed they were going to be intercepted on approach. CoCo would later describe it as a 'miracle' that they were not.

As they drew close, Glover began to livestream on his phone. CoCo stayed silent as he addressed the country. He might be a comedian, but this was no joke, he explained. In short succession, he laid out what he was doing, and why, and then explained Fireproof Australia's key demands: the creation of a sovereign aerial firefighting fleet, a program to rapidly rehome people who lost their livelihoods in catastrophic environmental disasters, and the smoke-proofing of kindergartens, schools and aged care facilities to protect children and the elderly.

'That's a call on the New South Wales government, it's a call on the federal government for an aerial firefighting fleet,' Glover said. 'They currently give $11 billion to the fossil fuel industry in subsidies, which is basically pouring petrol on the fire.

'Fireproof Australia is here now, about to conduct this public action. And I can tell you we don't want to do what we're having to do today. It's unpleasant, it's painful, but unfortunately it has to be done, because we're not getting the action that's required.'

At that moment, CoCo parked the Hino delivery truck, and those inside unbuckled their seatbelts. As she climbed out, she left the keys in the ignition to allow the police to easily move the vehicle from the road. From there, events moved swiftly.

CoCo would later describe entering an intense period of focus, like 'having the ball and trying to run down the football pitch'. Two other activists glued themselves onto the road in front of the van and held up a banner for Floodproof Australia, Fireproof Australia's sister activist group. Meanwhile, Glover and CoCo climbed onto the van's roof.

'All right, we're here,' Glover said, panning the camera to show where they were. 'We don't want to do it, we have to do it. There's four of us down here.'

At first, the scene was almost peaceful. Behind them, a familiar brick pylon rose out of the water and the bridge's iron rib cage curled away back towards the city. Looking around them, CoCo – who has a fear of heights – appeared almost calm. Inside, however, she was hyper-alert. It was 8.28 am when they arrived, and they had driven past two police cars parked on the bridge on the way over. Within minutes, an officer on a motorbike pulled up to take control of the scene, and Glover handed the phone to CoCo so she could take her turn to speak.

'People cannot get to work when their houses are flooded, they cannot get to work when there's bushfires, they cannot get to work when there's too much smoke,' she said.

As she spoke, some people driving past thanked them; others hurled abuse. A battery of sirens blared in the distance. As CoCo spoke, they grew louder and louder.

It was around then Glover lit the first flare.

'This is an emergency,' CoCo declared. 'This is an emergency, we are in an emergency situation. And we have to act now.'

More speeches followed as law enforcement arrived. Police diverted a truck caught around their van to clear the lane. After another few minutes, CoCo noticed the news helicopter circling. Fireproof Australia had its own media team, but after several weeks of non-stop activity they were burnt out and the group on the bridge was relying on legacy media to spread the word. To get the attention of the helicopter, CoCo lit her own flare. Passing the phone to Glover, she held the flare above her head like a torch as she looked out across the water. In the backdrop was Circular Quay and the Sydney city skyline. She held the pose as Glover recorded what would become the defining image of the protest. It was then the police began to make their approach.

'Here they come,' Glover said.

The response was instant fury. Not only was the Sydney Harbour Bridge synonymous with Australia, but the activists had struck a parochial sore spot. For those living on the North Shore, the bridge acted as the morning thoroughfare on their commute to work in the Sydney central business district. Traditionally a Liberal bastion in both state and federal politics, the area had also been the political stalking ground for former prime minister Tony Abbott until he was dethroned in 2019 by the climate-minded independent Zali Steggall. The preferred way in or out was to drive. By targeting the bridge, the activists had inadvertently targeted Sydney's professional managerial class. The result was that a minor disruption to their morning commute was met with incoherent anger.

Talkback radio erupted into condemnation, as did a Liberal state government nervous about looking weak ahead of an upcoming state election. The moment would also prove a test for the Perrottet government's new anti-protest laws that had been passed just 13 days earlier. These ratcheted up offences for 'disruptive' protest and their passing happened to coincide with the passage of similar laws in the US and the UK.

Over in the UK, the Conservative Party had pushed through its own laws in response to the activities of groups like Just Stop Oil and Insulate Britain. In the US, a similar crackdown had begun in response to the 2016 Standing Rock protests. There, the Sioux nation rallied hundreds against the construction of an oil pipeline that ran beneath a river just outside their reservation in North Dakota. They were met with water cannons and rubber bullets.

The legacy of Standing Rock and the forming of Extinction Rebellion so terrified oil and gas companies that their allies in conservative politics soon got busy rewriting the rules to stop them. In the US, this work was handled by the American Legislative

Exchange Council, a group of activist conservative lawmakers who specialised in writing model legislation that could be easily shared around the country and passed into law. As climate protests began targeting fossil fuel companies and their infrastructure, they set about writing model bills to 'protect critical infrastructure' that could be quickly picked up by state legislatures. By 2022, these bills would had been adopted in 24 US states.

Around the same time, over in the UK, right-wing think tank Policy Exchange began agitating for a massive expansion of police powers to crack down on protesters. With former Australian foreign minister Alexander Downer as chairman of trustees, the organisation was a go-to policy shop for the political right – with a record of having taken oil industry money. In 2019, it published what it called a 'review of ideology and tactics' of Extinction Rebellion, titled *Extremism Rebellion*. Three years later, British prime minister Rishi Sunak would be ratcheting up penalties faced by climate protesters. He would later thank the organisation for having 'helped us draft' the laws during a speech at a think tank event in June 2023.

In New South Wales, the state government's own laws appeared to be a variation on those in the UK. Under the new legislation, those found guilty of protesting illegally on public roads, rail lines, tunnels, bridges or industrial estates risked fines of up to $22,000 and jail for a maximum of two years. The wording was so broad that a bigger-than-expected gathering which spilled out across the entrance of a train station could potentially have been considered a crime.

When it came down to the vote, the New South Wales Labor opposition supported the Perrottet government and the bill was passed. Though Labor had initially expressed hesitation, it feared being wedged ahead of a state election and gave its support when a carve-out was negotiated specifically to protect union-organised rallies. Afterwards, Labor leader Chris Minns described the laws as necessary to protect the 'safety and security' of New South Wales.

Among the MPs who voted to pass the bill was Violet CoCo's uncle, a senior minister in the Perrottet cabinet. He would later respond to questions about his support for the bill, saying simply: 'nobody is above the law.'

Although the Sydney Harbour Bridge protest took place in mid-April, it wouldn't be until December that CoCo would get her day in court. The morning she arrived at the Downing Centre Local Court, things would not go well. Another magistrate had initially been assigned the matter, but a last-minute roster change delivered Magistrate Allison Hawkins.

Because she was a high-profile figure, police prosecutors threw every charge they could at CoCo, including assaulting police and, for firing off the flare, unauthorised use of explosives. CoCo pleaded guilty to several charges, but during sentencing Hawkins leaned heavily on police accounts. The prosecutor had piled it on thick, pointing to CoCo's long arrest record for similar protests across the country. But it was the ambulance that did it. By shutting down morning traffic on the Harbour Bridge, police alleged, the activists had trapped an ambulance responding to an emergency in gridlock.

Only there had been no ambulance. Footage from the event clearly showed the protest had blocked just a single lane of traffic, and any jam had been resolved in minutes thanks to the swift response by police – who provided no evidence to support their claim. Magistrate Hawkins, however, waved off any challenge to this suggestion. In sentencing, she latched onto the image of the ambulance, lights ablaze and siren screaming, helplessly stuck in traffic, unable to assist someone in crisis. The magistrate scolded CoCo, calling her 'childish' and 'emotional', adding that she was not a political prisoner but a criminal. She sentenced CoCo to 15 months' jail with an eight-month minimum.

'You have halted an ambulance under light and siren,' Hawkins said. 'What about the person in there? What about that person and their family? What do they think of you and your cause?'

Hawkins' remarks in the courtroom would be widely reported, entering the public imagination and shaping the narrative around what happened. It didn't matter that the Fireproof Australia activists ensured access to emergency vehicles in their actions as a matter of policy, the motif made for a politically useful myth. The purely hypothetical ambulance would be used to justify both the New South Wales anti-protest laws and those in other jurisdictions.

CoCo's lawyers applied for bail, but the application was summarily refused by Hawkins. The time of year meant the decision would have resulted in CoCo spending Christmas in a cell had she not secured a fast-tracked hearing in a higher court. Looking back, CoCo would say she believed Hawkins knew this and that the magistrate's decision was intended to punish her.

Outside the courtroom, news of the sentence caught the attention of Clément Voule, UN special rapporteur on the rights to freedom of peaceful assembly and of association. CoCo wasn't the first to have been hit with a severe sentence: a 22-year-old Blockade Australia protester from Newcastle had been sentenced to 12 months the year before. In a social media post, Voule said he was 'alarmed' by CoCo's sentence, as 'peaceful protesters should never be criminalised or imprisoned'.

Dominic Perrottet, then New South Wales premier, disagreed. Instead, he took the opportunity to gloat about his government's victory on talkback radio.

'If protesters want to put our way of life at risk, then they should have the book thrown at them, and that's pleasing to see,' Perrottet said. 'We want people to be able to protest but do it in a way that doesn't inconvenience people right across New South Wales.'

She chose the location for the poetry if nothing else. Standing in the East Lismore General Cemetery at the crack of dawn, on the one-year anniversary of the 2022 flood that had killed five Lismore residents and displaced thousands, Violet CoCo pointed to a neat row of houses that had been inundated during the disaster and started to describe what had happened.

'On this day, one year ago today, people all through here and the rest of Lismore were being evacuated from their homes,' she said.

'It still looks like a war zone here. There are six thousand people who a year ago lost their homes to the floods in the Northern Rivers, and not a single person has been cared for by the government yet in that respect.'

CoCo was weeks out from an appeal's hearing to challenge her eight-month minimum sentence and had already endured heavy-handed bail conditions that kept her under a strict curfew. It was yet another tactic normally reserved for organised crime figures that had been applied to climate protesters across the country with increasingly frequency. These restrictions kept the troublesome activists at home and stopped them from seeing their friends, or in some cases, their family. It also gave individual police officers total discretion over their future.

CoCo had initially been slapped with a strict 24-hour home detention order. Police could rock up at her door at any time of day or night; if she failed to respond, a warrant would be issued for her arrest. If she slept through a 3 am doorknock, for instance, she would find herself back in cuffs, serving out her full sentence.

She had moved to Lismore to help with reconstruction, but also to escape the intense police monitoring. It was 'a push-pull scenario'. She wanted to help rebuild, and she would spend considerable time volunteering to remove rubbish and carrying out basic construction work. She also reasoned it might persuade police to drop the curfew. In the end, they succeeded in running her out of town but only shortened the curfew to apply within certain hours of the morning and evening.

By this time her situation had become emblematic for what was beginning to look like a campaign of repression targeting climate activists across Australia. Having failed to meaningfully address climate change for two decades, Australian authorities were responding to protest movements demanding they do more with force.

First to move was Queensland, whose Labor government passed anti-protest laws to ban 'locking devices', based on false claims that climate activists were using them in their campaign against the Adani coalmine. Then came New South Wales, Tasmania and South Australia.

The Western Australian Labor government took a different tack when it defaulted to force. The first signs occurred in August 2021 when six Extinction Rebellion activists protested the $16-billion Scarborough gas development in the north of the state by painting anti-Woodside messages on a footbridge within view of company headquarters. There had been nothing out of the ordinary about this activity. The chalk paint was washable, meaning it wouldn't cause any permanent damage and no-one blocked access on the bridge itself. Police had even been present to keep an eye on their activities. Afterwards, everyone went home with no issues or complaints.

Two weeks later, officers with the State Security Investigation Group – a division of the Western Australian police force commonly referred to as the state's 'counter-terrorism police' – raided the homes of all six activists in the early hours of the morning. As a shadowy body within Western Australian law enforcement, their existence is not advertised. There is no public budget item that lists their funding and their work is entirely exempt from freedom of information requests. According to its operational charter, the group's responsibilities include targeting 'political, racial or ideological motivation extremist unlawful behaviour' and 'issue motivated groups who conduct activities which may lead to violence'. The precise criteria used to identify and target any given organisation are unknown, but the group worked in cooperation with the Fixated Person's Unit.

Four of the Extinction Rebellion activists would cop penalties when they pleaded guilty to damaging property, and two would have the charges dropped when they fought them in court. Later, the same police unit would appear to target those protesting the threat to ancient rock art on the Burrup Peninsula by industrial development, including a massive new gas processing hub proposed by Woodside. The reason seemed to be an overlap between the groups. In another incident, activists used stink bombs to force an evacuation of Woodside's 2023 annual general meeting.

Woodside responded as if it were under siege. The company posted a permanent 'statement on activism' to its website following the protest at its AGM, calling for 'these actions to be met with the full force of the law'.

Similar tactics would also be adopted in New South Wales. When activists from Blockade Australia targeted the Port of Newcastle in December 2021, the state government formed a strikeforce – a police taskforce normally reserved for organised crime – to actively watch and disrupt the disruptors. In their efforts to arrest the radicals, police ended up raiding the Hunter Environment Centre which had nothing to do with organising the protests, hauling away their computer equipment for forensic examination. The next year in June, Blockade Australia was again raided after activists camping on private property stumbled upon two people in camouflage spying on them. The two men turned out to be police and after claiming they were set upon by protesters, the campsite was raided. The scope of these activities would broaden significantly in October that year. When planning its annual event, the International Mining and Resources Conference appeared to shift the location from Melbourne to Sydney to take advantage of the state's anti-protest laws – an exercise in jurisdiction-shopping.

Ahead of the conference – speakers included representatives from governments as far away as Saudi Arabia, Sudan and South Australia – New South Wales police asked its sister agencies to pre-emptively visit a list of known environmental activists, many of them interstate, on the off-chance they might be protesting. Ostensibly this effort was to explain their rights as protesters, but many of those contacted were not involved in any campaign against the conference and many had no real affiliation with the more militant groups. Those who received a visit reported being greeted by police officers at their front door. Often, the officers read from scripts explaining how they would need to fill out a specific form to protest the conference in New South Wales. None of this had a basis in law, but it doubled as a means of intelligence-gathering. Any information people offered unprompted could be fed back to New South Wales law enforcement.

The trial of Violet CoCo became a symbol for this broader crackdown. For the authorities, her appeal represented a test of their righteousness and strength, but CoCo had just as much at stake. Failure meant prison. Publicly she would say a jail sentence would be a necessary sacrifice for the cause, but in more reflective moments she confessed to being apprehensive.

'Politics is theatre,' she said. 'Directing a nonviolent direct action is like directing a theatre show.'

'You have all your roles, and everyone's got a script. You're going to bump in and bump out, but I guess the major difference is the guns are real.'

When her day in court came, a crowd had gathered in the small public terrace out front of the Downing Centre. Young radicals and old hippies filtered in to join the solidarity rally on the crisp golden morning in Sydney. As the crowd grew, police arrived to ensure no-one stepped onto the road. Violet CoCo sat across the street at a coffee shop table with her lawyer, Eddie Lloyd, watching the scene unfold. Under her bail conditions, CoCo was not allowed to set foot in Sydney until the day of her hearing. This made getting to court from Lismore, an eight-hour drive, logistically difficult. To avoid failing to turn up for her hearing, or breaching her bail conditions, she had driven down the night before and camped at Freemans Camp Ground, one-and-a-half hours up the coast.

There was speculation among the crowd that CoCo might speak – another potential violation of her bail conditions. When she did not, others took up the responsibility. Passionate addresses were made about democracy to those assembled, and then the crowd began to march. Circling the courts, they chanted slogans until it was time for CoCo to make her way to court. As she made her way over, the crowd marched past a final time. CoCo raised her fist in solidarity.

Inside the courtroom, the atmosphere was clinical. The public gallery filled as media, supporters and lawyers filtered in. District Court judge Mark Williams would oversee the matter and the softly spoken man was in no mood for a circus. In a hearing that lasted roughly an hour, Williams gave short shrift to efforts by crown prosecutor Isabella Maxwell-Williams to paint CoCo's offending as a 'complete disregard for public safety and the rule of law'. Footage recorded by Channel Seven of the protest played for the court was set aside. Police claimed this video showed that the disruption had caused a major traffic jam, but Williams responded with a shrug. It just looked like morning traffic in Sydney on any given day of the week. Another assertion that CoCo had been motivated more by scorn at a heavy sentence handed to her partner in a separate protest rather than climate change was also rejected. So, too, was the submission that the protest was 'not peaceful' or that CoCo should be punished harshly based on her 'criminal history', including having once spray-painted the Extinction Rebellion logo on the front window of the APPEA offices.

But once again it was the ambulance that caused the biggest ruckus. At trial, the New South Wales Police case hinged on the presence of an ambulance under lights and sirens, but in the lead-up to the appeal, the Crown had been forced to back down. During the trial of one of the other activists, the Crown admitted it had no evidence to support its claims and the ambulance did not exist. In their eagerness to secure a prosecution, police had embellished their account, and this material was redacted from the agreed-upon facts before the judge. When Williams was made aware of why it was redacted and how CoCo had initially been sentenced largely on this 'false factual basis', the credibility of the Crown collapsed. Williams was read a section of transcript from Magistrate Hawkins' original sentencing which hinged on this 'false fact'. Reviewing this evidence, the judge asked, 'How did that find its way in?'

'Your honour, that's something that the police will have to fess up to,' he was told from the bar table. 'That magistrate was misled.'

At that point, the tone of the hearing changed. What began with mild annoyance ended with clear frustration at what looked like a set-up that had wasted the court's time and misled a magistrate. When the hearing was over, Williams upheld CoCo's appeal and quashed her sentence. Convictions were recorded for blocking the bridge, resisting arrest and using a flare, but no jail time was imposed. CoCo was given a 12-month conditional release order.

'I don't accept the Crown submission that she is a danger to the community,' Williams said, flatly.

Violet CoCo, head in hands, cried with relief.

★

If the activists were increasingly coming under pressure, the shrill tone of APPEA press releases suggested that so too were the nation's oil and gas producers. They appeared in my inbox with increasing regularity in the lead-up to Christmas 2022 and read like Rex Connor himself had risen from the grave:

APPEA REQUESTS URGENT MEETING WITH PM AS MORE DAMAGING INTERVENTION EMERGES

GAS MARKET INTERVENTION A WARNING FOR AUSTRALIAN INDUSTRY

INVESTMENT RISK OF INTERVENTION SHOWN IN SENEX ENERGY PAUSE ON $1BN EXPANSION

Upon taking office, the newly elected Albanese government had broken with decades of hands-off, free-market orthodoxy, intervening to cap the price of gas to avoid a drastic increase right before Christmas. European countries heading into winter had turned to the spot market to fill their tanks and avoid funding the Russian war machine after it had invaded Ukraine. As Europe – Germany in particular – soaked up the world's gas supply like a sponge, the price had risen, and the same spikes seen elsewhere were looking like they would be repeated in Australia.

The price cap sent the nation's gas producers into a panic. At the same time, the government was also talking about fiddling with the safeguard mechanism – a complicated bureaucratic instrument superficially designed to lower emissions among heavy emitters. The gas producers responded as if they were being lined up against a wall.

APPEA, however, was in no position to push back. Cracks had begun to show in the organisation in 2021, when, a month after the International Energy Agency released its bombshell report recommending no new investment in coal, oil and gas, the association's CEO, Andrew McConville, shed tears during his closing speech at that year's conference. He had been thanking his staff for their dedication during the early pandemic period, but close observers noted the timeline and suspected he was feeling the pressure.

McConville had always seemed an odd choice for the organisation. A country boy, he had initially served with the Australian Wheat Board as a junior officer in charge of government relations when the oil-for-wheat scandal broke. It had been a dark chapter in corporate Australia's history. Senior officers at the organisation had engineered an elaborate workaround to evade sanctions imposed on Saddam Hussein's regime that involved the payment of kickbacks to the Iraqi government. When it was discovered, all hell broke loose. As a junior officer, McConville was not directly responsible, but he was among those who gave evidence to the ensuing royal commission, at which he claimed he had lost his notebooks. After this experience he left the country for a stretch, taking a job overseas in government relations for Syngenta, a Monsanto analogue that lacked the same name-recognition. He made his way back to Australia when APPEA brought him on board as their CEO in April 2019. McConville would make his exit during the dying days of the Morrison government, when he was offered a job heading up the Murray–Darling Basin Authority, forcing APPEA to scramble for a replacement.

Having assembled a hiring committee, they decided their preferred candidate would be a woman with environmental credentials and an engineering background, but some of the first women they approached knocked them back flat. Looking overseas, they settled on Samantha McCulloch, who was then working for the International Energy Agency, heading up a team promoting carbon capture and storage. McCulloch was a technocrat who had previously worked as a policy wonk for the Department of Industry, the Department of Resources, Energy and Tourism and the Australian Coal Association. Her new brief with APPEA would be to sell gas as a clean fuel essential to any transition – the same line the industry had been running since 1973.

By the time she took the job, a staff exodus was already underway. Sarah Browne, regarded as a highly effective public affairs manager who had steered the association towards a less combative, middle-of-the-road messaging, took a job with the Port of Melbourne. Northern Territory director Cassandra Schmidt also made an exit, as did the Queensland director, Matthew Paull, and the Western Australia and South Australia director, Claire Wilkinson. Senior policy adviser Luke Earnshaw took a role with the Department of Industry, Science and Resources, but the most high-profile loss was deputy CEO Damien Dwyer, who ended a 17-year career with the association – though he still maintains a position on the AIGN board. When Dwyer walked, so did a wealth of institutional knowledge and history.

Perhaps the most critical loss, though, was Ashley Wells. Wells had previously advised Labor's Kevin Rudd and Stephen Smith before taking up a role with APPEA as its government relations manager. He left to hang out his own shingle, setting up a public relations consultancy that picked up contracts from APPEA members Beach Energy and Esso Australia – a sign of their wavering faith in the industry association. Other, more junior staff also quit, and the loss of muscle left APPEA – and the broader oil and gas sector – at its weakest in generations.

The displeasure among its member companies was self-evident when I joined 150 gas producers at the Sheraton Grand in Sydney in March 2023 for the 10th annual Australian Domestic Gas Outlook conference.

For the most part, it was more of the same for a gathering of fossil fuel producers, except this one involved just the gas people. Beneath the froth, there stirred a deep pool of despair. At lunch, they dined on piles of oysters at the buffet tables, but during the sessions, they muttered about how companies should pull their investments from the country as a show of force to protest the government's gas price cap. The execs offered up all the usual grievances, complaining that they were unappreciated by the public.

Everyone knew times were tough. Strategy documents published on the International Gas Union's website had already warned its member organisations that action against climate change was 'potentially existential for the global natural gas value chain'. The International Gas Union was a global equivalent of APPEA but for the gas industry. In an effort to help, it had drawn up region-specific messaging plans that opportunistically leveraged whatever was happening locally to defend gas.

These talking points had already cropped up in APPEA's own comms, which sought to recast gas producers as heroes of the transition – but among this crowd, it wasn't enough. During questions, a member of the audience informed the panel that industry associations in Canada 'rigorously defend the development of resources' against 'significant community or social opposition', lamenting the fact that this wasn't so in Australia.

I wasn't the only one who had noticed the cloud of depression settling over the conference. Sitting in the audience, Misha Zelinsky had also been taking careful note. He had been booked as a speaker to talk about the Russian invasion of Ukraine but had turned up early to take the temperature of the crowd.

It wouldn't the first time he had addressed this audience, but as an ex-union figure, he did seem out of place in this room. A lawyer, Zelinsky was a Fulbright scholar who had worked up an impressive résumé. Formerly assistant national secretary of the Australian Workers' Union and secretary of the Australian Labor Party's National Policy Forum, he had been well on his way to securing pre-selection as a Labor MP. It was not to be, however, thanks largely to a self-inflicted wound. As a younger man, Zelinsky had co-authored a book titled *He's an Arsehole Anyway,* written in the kind of back-slapping, shit-stirring tone of a best man at a wedding. It was a satirical work, written in the voice of 'the arseholes themselves', offering misogynist advice to women who had recently been dumped – in a way that was no longer tolerated post #MeToo. Having failed to pass a vetting committee, he was forced to pull out in 2022. Undeterred, Zelinsky reinvented himself as a foreign correspondent when he travelled to Ukraine following Russia's invasion, filing dispatches to the *Australian Financial Review.*

Zelinsky had been invited to the conference at the Sheraton to speak about the war, but as he got started it was clear he had no intention of following through. With some purple rhetoric, he paid tribute to the struggle of the Ukrainian people against their aggressors before pivoting. The real subject he wanted to discuss was the 'no friends' problem faced by gas producers.

'If I could be this blunt to you all, Mr and Mrs Gas People,' he said, 'do you have any friends? I mean, serious, actual friends?'

Between jokes and compliments, Zelinsky explained how, through its own hubris, the industry had successfully managed to annoy, upset or anger just about every potential friend it had over the last two decades. He explained that the average Australian saw them as 'tax-evading, union-busting, manufacturing-destroying, family-gouging, foreign-raiding, climate-vandalising, cold-hearted war profiteers' – and he dared them to poll it. Gas had gone from a 'bridging fuel' to 'just as bad as coal' in record time, he said, and it had been entirely predictable.

The activists' playbook, Zelinsky told them, had three critical steps. 'Step one is create a causal link between a big problem and an industrial process,' he said. 'Step two: stop expansion. Once you've stopped expansion, you've now delegitimised the industry. Now, once you've got the net in place, close the net, as step three, on existing production.'

According to Zelinsky, activists had done this with coal, which had once been thought untouchable, and now they were coming for gas producers. Later, while answering audience questions, he would say there was no way to cut a deal with the 'hardliners' among the activists – which was broadly true. Radicals had a tendency to splinter and reform over questions of ideological purity, tactics or just interpersonal spite. Zelinsky's suggestion was that industry might try reframing future negotiations on friendlier terms.

But he was also explicit in his prognosis. The industry didn't have an image problem, Zelinsky explained – though they did. They didn't have a messaging problem, though they also had one of those. Their real problem, he said, was a crisis of legitimacy. Through their actions, the gas producers were accelerating the creation of a world where people no longer needed them – and everyone listening knew it. Gas producers once had to work to convince Australians to switch from town gas, made from coal, to using what they pulled from the ground. Not only did they have to replace or retrofit every stove, heater and hot water service – there were 400,000 appliances among Adelaide's 150,000 gas customers alone – but they also had to teach people to embrace gas as a 'cleaner' fuel. Once, the Australian Gas Association even recruited a home economist to offer educational sessions and cooking tips to housewives. Documentary films were made about the benefits of gas, and on 20 March 1967, the *Australian Financial Review* ran a 28-page feature story on the transition.

Half a century later, there was now a growing movement to phase out gas by banning connections in new homes. Victoria had already gone down that path, and there was an expectation that, eventually, other states would do likewise. The reality was, Zelinsky reminded them, 'the more removed [gas] becomes from the consumer and the public and the voter, it's more dangerous for you'.

'If you're seen to be illegitimate, and your good is potentially – potentially, at least – substitutable, you no longer have a guaranteed licence to operate,' Zelinsky said. 'It can happen really quickly. Why else do you think the Saudis are paying Greg Norman billions of dollars to run around making a goose of himself, right? Not because they suddenly love golf, but because they can see what's coming down the pipeline.'

Having made his point, Zelinsky then moved to close. He gave a quick run-down of the parliamentary numbers, pointing out that 'despite all the noise about Teals and Greens', those who wanted change 'don't have the whip hand politically – yet'. What he proposed was a strategy of containment – a chance to 'strike a bargain' between Labor and the Coalition on the future of the gas industry to head off the Greens.

'It would also be handy, in my view – and this is just free advice – to have a loud, powerful voice in the room making decisions,' he said.

One such friend who might help them out, Zelinsky intimated, was the Australian Workers' Union, which regularly sent representatives to industry events, including this one. The AWU membership was 'acutely aware their industries are powered by gas' and had previously worked with the Tasmanian salmon industry to turn around public perceptions in the face of community opposition over environmental concerns. There was already a raft of federal legislation on its way, he pointed out, listing off each new reform in turn to make his message clear.

'The government's actually going to need as much help with this as you do,' he said.

The gas producers might still be indignant about the price cap and 'choose to go to war', but he reminded them that 'wars are costly'.

'They're unpredictable, and they're often more crippling to the side waging it,' he said, in an allusion to Russia's invasion of Ukraine.

I looked around the room to gauge the reaction of those listening. The mood was one of quiet awe. At the very moment they were at their weakest, Zelinsky was offering redemption. Not only had he diagnosed their condition, he was offering them a hand up – with a charisma that cut through the stifling, myopic self-pity. In half an hour, he had outlined a coherent political program, and he hadn't charged them a cent in consulting fees.

In the weeks that followed, I kept an eye on APPEA's press releases and soon noticed a change in tenor. A tense suspicion remained, but it seemed a new arrangement had been struck. Gone were the panicked threats of an investment strike, in favour of more conciliatory language reminding people what the industry was worth – and that it must continue:

GAS SECTOR TO DELIVER $16 BILLION TO GOVERNMENT AS CONTRIBUTION ALMOST TRIPLES

BEETALOO GREEN LIGHT USHERS IN ECONOMIC PROSPERITY AND EAST COAST ENERGY SECURITY

FUTURE ROLE OF GAS AND NET ZERO TECH HEADLINE FEDERAL BUDGET UNDERPINNED BY OIL AND GAS CONTRIBUTION

15

Government Is at Your Disposal

There was a nervous tension among the attendees at the 2023 APPEA conference. That year the show would be held in Adelaide over three days from 15 May. Gone was the buoyant hubris of Brisbane, just after Russia's invasion of Ukraine had T-boned growing political momentum for climate action, filling the petroleum industry's coffers on the back of rising prices. This time around, the vibe was flat, as if the air had been sucked right out of the Adelaide Convention Centre and everyone was holding their breath.

Part of the reason was the security threat. The event organisers had clearly been briefed and were taking no risks. An unknown number of police kept a discreet watch over the exterior of the conference. With most of the entrances closed off, just getting inside from North Terrace required heading up an escalator and through a makeshift security checkpoint in a central plaza. There were two stations at this juncture, where attendees' bags were checked, and they were eyeballed by security for their official lanyard. For good measure, three burly, neatly dressed security guards stood in delta formation just inside the entrance in case anyone needed to be tackled to the floor.

Everyone knew Extinction Rebellion was coming. The group had been planning for the conference several months out and hadn't been shy about its intentions. Initially there had been talk about a mass mobilisation of activists converging on Adelaide, but when the time came this plan seemed to have been substantially scaled back. I never found out why, and the industry people didn't seem to notice; for them, one protester was one too many.

Much had changed in the year since the petroleum people last met. The industry had made multiple billions in profits thanks to a war of aggression carried out by a petrostate – Russian gas giant Gazprom had even deployed its own combat militias ('Steam' and 'Torch'), recruited from among its employees, as part of the invasion force. But the endless stacks of cash that Australia's oil and gas companies were racking up had brought greater scrutiny that was fast evolving into public calls for a windfall tax. Worse, APPEA had been powerless to stop the government intervening to cap gas prices – and then there was the ultimate indignity of all those people comparing petroleum companies to Big Tobacco in the press.

Spooked by the public backlash, the new APPEA CEO, Samantha McCulloch, was doing her best to salvage the gas industry's reputation. Ahead of the conference, APPEA had published a preliminary document titled 'A review of net zero energy and industrial zones', written with technical support from the CSIRO. This review included a map of the Australian continent highlighting nine future 'net zero' industrial zones across the country, all grouped around existing gas infrastructure.

Among them was the entire region of the Pilbara in Western Australia, and also the proposed Middle Arm development outside Darwin in the Northern Territory. The Territory government had plans to turn a pristine marshland peninsula into a concrete slab that would in theory become home to a renewable energy export industry. The government, however, had been coy about who would be setting up shop on site. As it later turned out, a government video presentation obtained by the Environment Centre NT under freedom of information laws showed one of the biggest operations on site would be a petrochemical plant run by Saudi Aramco, the world's largest oil company. Other government documents showed that the project itself would require damming the Territory's

Adelaide River to ensure a constant supply of water. The Northern Territory, for what it's worth, is one of the last places in the world where all its rivers are free-flowing.

McCulloch was calling for the creation of what's known by academics as 'industrial clusters', groups of complex industrial operations built in close proximity to increase their efficiency and collaboration. These were supposed to set up an ever-improving cycle of capability growth thanks to short supply chains and innovation sharing. To communicate this idea, her team had come up with a simple analogy. 'Carpooling' industrial activity in this way, she said, would help Australia achieve net zero 'in the fastest, most cost-efficient way'. It would also lock in the long-term survival of the nation's gas producers.

'Leveraging existing infrastructure can ensure regions that powered the Australian economy to where it is today won't be left behind as we restructure our economy for the future,' McCulloch said.

McCulloch would briefly throw the report's map up onto the screen during her opening address as a visual representation of the necessity of oil and gas. Just getting to this point, however, seemed to have been a struggle for the organisers of the event. The annual golf tournament was postponed for another year, 'due to committee and course availability'. Hire-in labour also came at a premium. Before the event, I got talking to a camera operator who explained how, when approached by the event organisers, she had quoted her regular fee. The organisers asked whether she was sure about the amount and countered with an offer three times as much, telling her to think about it overnight. After she signed, they explained she could have asked for anything, and they would have paid it.

'Danger money,' she said, raising an eyebrow.

Even the guests were picking up on the vibe. Federal resources minister Madeleine King, speaking to the conference for the first time since the election of the federal Labor government, opened the show. She kicked things off with all the same tired jokes about how protesters outside had found creative new uses for the petrochemicals that made up superglue. She assured the industry that it had the government's support – but it was the following speech from South Australian minister for energy and mining Tom Koutsantonis that stole the show.

There was no hint about what was about to happen as the minister stepped up to the microphone. Koutsantonis has always had a flair for flair, but his speech that morning would prove to be something else. What began with a preamble thanking his hosts quickly transitioned into a heartfelt expression of gratitude to the industry:

A lot of people in South Australia don't know the importance and the impacts of the resources industry on our state. The untold wealth that this industry has given to our state has been, to date, $3.6 billion in royalties to this state to build hospitals, to build schools and industrialise our state. It gave us a standard of living that is second to none. Across the country. We are grateful for the work that you have done. And it must be difficult – these last few months in your industry, you must feel like you're under siege. It must feel as if you're not welcome. It must feel as if the work, the endeavour, the innovation, the risks that you all take are somehow not thought of as productive for our country. Nothing could be further from the truth. Nothing could be further from the truth. This industry, as Madeleine said, is a key pillar in our path of decarbonisation. We cannot do it without you. We cannot make advancements without you. We cannot have net zero without you. I cannot decarbonise our steel industry without this industry. We can't decarbonise our electricity industry without this industry. We cannot transform our economy to net zero without you. You are key partners in this endeavour. And those people who are misinformed outside will tend to belittle or criticise this industry and do not understand the depths of the innovation that your industry is making to help this nation and the world to decarbonise.

It would be an understatement to describe the minister's opening remarks as 'over the top'. There are few words in the English language capable of capturing how bizarre it was to watch a senior cabinet minister cooing over a room full of oil and gas executives like they were a child with a skinned knee. But that was not all. After a 15-minute speech gushing about the importance of gas, the minister went even further, closing out his welcome with a display of deference.

'We are thankful you are here,' he said. 'We are happy to be a recipient of APPEA's largesse in the form of coming here more often.

'The South Australian government is at your disposal; we are here to help, and we are here to offer you a pathway to the future.'

A couple hundred pairs of hands broke into a rapturous applause and I did a double take. He had offered no caveat, made no attempt to qualify, no effort to preserve the independence of his office. He had not only put the state government 'at the disposal' of the petroleum sector but was guaranteeing its long-term survival.

Over the course of the conference, the moment would be brought up again and again. Santos managing director Kevin Gallagher would describe it as a 'breath of fresh air'. Others would fawn, saying they wished for a similar welcome at every event.

It was day two when Extinction Rebellion struck. At the agreed-upon time, a group of activists closed the road beneath Morphett Street Bridge and flew into action. Meme Thorne, 69, secured her harness, climbed over the railing on the eastern side and lowered herself down. Wearing a shirt that said 'NO NEW COAL, OIL OR GAS!' she hung there, suspended in midair, blocking east-bound traffic along North Terrace. Though both the tramway and the west-bound lanes remained open to emergency vehicles, any other car travelling that route would be forced to redirect through other parts of the city.

At that time of morning, the result was gridlock across Adelaide's northwest corner. I was sitting in my car around the corner, wondering why I was stuck in traffic, when it happened. I found a park and walked over just in time to catch the first police responding to the incident as they arrived. Taking in the scene, the lone officer looked genuinely befuddled, overwhelmed by the logistical challenge of getting Thorne down.

What the conference had come down to was duelling spectacles staged by the oil and gas executives and the activists. Inside the walled garden of the Adelaide Convention Centre, APPEA was working hard to project an aura of quiet confidence, certainty and consensus in front of the nation's financial press. Outside on the street, Extinction Rebellion was mounting a direct challenge to this display with its own, highlighting the role of oil and gas in driving the existential threat of climate change. The protest at Morphett Street Bridge would be one of several to take place over the course of the conference. Fresh from an arrest in Perth, Violet CoCo herself would even fly in to join a sit-in at the South Australia Drill Core Reference Library, where she'd be arrested once more.

If their duelling ground was the media, the protest on the bridge would prove the most provocative manoeuvre by the activists. There was nothing new about this sort of protest. Similar actions had been carried out in other, distant parts of the country over the past decade – but now that it was happening in Adelaide, it would stir up a good, old-fashioned moral panic.

South Australian police commissioner Grant Stevens helped get it started when he addressed reporters during a press conference following the protest. According to Stevens, there was supposed to be a proper order and process to these things. The audacity and defiance of the action that morning had apparently made him so angry that he was contemplating grievous bodily harm.

'We can't, as much as we might like to, cut the rope and let them drop,' he said.

This was the voice of authority, captured in a perfect soundbite. Almost as soon as these words were reported, talkback radio went to work stoking people's rage, and political leaders who might otherwise mock 'government by Twitter' began paying close attention.

Little of this penetrated the conference. Inside the convention centre, the vibe was serene. Well-credentialed experts sat on panels and gave talks with titles like 'New Energy Technologies: Opportunities and Realities', 'Lead, Shape, Innovate: Accelerating to Net Zero' and 'Protecting Our People'. The only acknowledgement of what was taking place outside came in the snide jokes panel participants shared on stage.

Resolute and relentless, the show rolled on.

It was literally a circus. The gala dinner that year was themed 'May Soiree', and considerable effort had been made by the organisers to look as if they had gone to considerable effort. Tickets were $205 a pop, and this time I had not been given a freebie, meaning I paid my own way. Looking to recoup this expense, I was among the early birds turning up slightly ahead of time to get nicely toasted before the party started.

For the attendees the party would be a chance to break away from reality for a while, to forget the gauche security checkpoints and killjoy protesters, to duck the flat mood of the conference and let loose. A waiting area had been loosely roped off for service of alcohol in a sterile hallway on the west side of the convention centre with no view of the Torrens River, likely chosen to avoid a raid by activists. Two stilt-walkers dressed like old-timey English toffs in waistcoats and top hats did the rounds among the crowd – a growing herd of men in blue and grey suits – before taking up a position by the entrance. They laughed and jigged and greeted new arrivals, welcoming them by repeating the word 'soiree' in ghostly, theatrical tones.

I found a spot by an information counter that had been appropriated as a drinks' stand and scrolled my phone. A story in *The Guardian* carried the headline 'World likely to breach 1.5C climate threshold by 2027, scientists warn'. The strap explained that the UN was warning that El Niño and human-induced climate breakdown could combine to push temperatures into 'uncharted territory'. I looked around the room. It didn't seem a message that would get much cut-through with this crowd.

That evening, as numbers outside the dinner hall swelled, the industry's heavy hitters filtered in. According to the seating board there were 87 tables, and my ticket told me I was on table 72. It was almost at the very back, right in the corner, and so far away from

the centre I might as well have been in Siberia. When the doors
swung open to let in the guests, the room was awash in pink light.
Colossal strips of fabric hung from the centre of the ceiling, creating
the impression we were in a circus tent. On the tables, bedazzled
lamps provided a sense of ostentatiousness. Hunting for my seat, I
eventually learned I would be joining half a dozen guys from the
same oil company, a smaller operator with no real public profile. On
my left was their chief financial officer. On my right was a long-term
veteran of the oil industry who had twice met Dick Cheney when
a company he was with in Texas did some work with Halliburton.
They were nervous when I explained I was a reporter and asked if I
was there for a story.

'Right now, I'm just here for dinner,' I said, and it was God's
honest truth.

I eyeballed the menu; the offering would have been enough to
make Marie Antoinette blush. Entree promised a 'taste of South
Australia' with 'Eyre Peninsula oyster, finger lime caviar, KIS vodka
and Geraldton wax-cured kingfish, prawn and blue swimmer crab
rillettes, lemon and caper remoulade, native herb oil'. Main was
beef tenderloin with 'saltbush hasselback potato, cauliflower and
native thyme puree, warrigal greens, blistered truss tomato, pickled
mustard seeds and a shiraz jus'. Desert was a lighter affair, with a
plate of 'Callebaut Gold namelaka, spiced sponges, whipped coconut
ganache and freeze-dried native fruits'. Accompanying each course
was an assortment of wines, with tea and coffee served in the lounge
afterwards.

By now I'd been to a few of these events. Along the way I had
learned how it took about two cups of diesel just to make the beef
patty in a burger. This was the amount of fuel needed to raise the
cow, move it, kill it, process it, ship it and portion it. I wondered
how much fuel it had taken to put together this menu for a couple
hundred people.

When the show began, it was clear there had been a few changes
made for this year's event. For starters, no-one read out a list of the
high-profile attendees to thank them for turning up. There were
also no long speeches during the awards.

But there was also another omission. The petroleum engineer on my right leaned past me to point out to his CFO on my left how no-one on stage had once uttered the word 'oil'. It had been the same throughout the conference. No mention of oil at an oil and gas conference seemed like a snub, he grizzled. If the organisers were so ashamed of the oil companies among their membership, what was the point of APPEA?

It turned out this was a question the association itself was asking, and in September 2023, it announced a rebrand. From that point forward, they would be known as 'Australian Energy Producers' or AEP – an initialism difficult to say as a word. When attempted, it sounded awkwardly like the word 'ape'.

Main courses arrived with the evening entertainment, a dreadlocked magician in a fedora who looked like a poor-man's Criss Angel. Between dance numbers, he performed various escape illusions with help from his assistant. Somehow it all seemed fitting.

By then the wine was flowing and the guys at my table had warmed up to me. We talked slow-cooking, dry rubs and Dick Cheney's habit of shooting his lawyers. It was only towards the end of the night that the CFO took a moment, protected by the din of the dancefloor, to ask me a burning question. Oil industry people live in an echo chamber, he explained. He wanted to know what the vibe was 'out there, about us'.

It was a question I was not prepared for. I think it was his uncertainty that caught me off guard. Over the past few days, the industry's corporate leadership had been working hard to project the illusion of confidence. Apparently, not everyone felt so certain.

By this point I had come to genuinely like the guy and didn't want to offend him, but the prognosis was clear. Petroleum companies were hated. People I spoke to generally described them as foreign-owned, tax-avoiding, union-busting, profiteering, climate-killing bastards who were making out like bandits over the price of petrol. The only people who defended them either worked for them, with them, or wanted to get in their pocket.

Worse, if we were to take climate change seriously, the oil, gas and coal industries were on borrowed time. At law a corporation was considered a legal person, and there were law professors already looking to extend the analogy to argue the world needed to put its oil companies into hospice. Faced with the prospect of a managed decline, companies could either accept the situation or they could fight it. Whatever they chose, a phase out had to be planned, even if no leader – corporate or political – was brave enough to acknowledge this reality.

Change was, of course, possible for the petroleum companies. There were plenty of examples out there. Finnish firm Nokia was founded in 1865 as a paper mill but a century later moved into the business of making mobile phones. So far as I was aware, no Australian oil company – let alone gas producer – was seriously reconsidering its core business. Hydrogen might prove an alternative but the executives kept complaining it was too expensive. Instead, they were cheaping out, eyeing off biogas and so-called 'e-methane' as possible alternative business lines. Shell had even brought out a selection of allegedly 'carbon-neutral' fuels which relied on accounting tricks to offset its emissions and claim a minimal environmental impact.

Sooner or later, however, these companies would be confronted with certain choices. They could either confront the issue head on or have the decision made for them.

On some level, everyone in the room knew this. They had to. The cynical were just trying to make it out with as much cash as possible before the music stopped. Meanwhile, the true believers among them seemed to be moving through the stages of grief but were stuck on bargaining. Until they reached acceptance, they were just lying to themselves.

Delicately, I explained to the CFO that I thought his was an industry stalled at a crossroads, unwilling to face up to its own illusions. Smaller operators might get away with continuing to work existing fields for some time, producing for increasingly niche applications, but the Exxons, the Shells and even the Woodsides of the world could not continue on as they had been. That was not a political statement; it was a reality of physics. There was a hard limit to the amount of carbon dioxide the atmosphere could absorb.

Trying to negotiate it was like trying to strike a bargain with gravity while in freefall.

The CFO said he was very concerned about overpopulation and asked how it might be possible to feed everyone without oil. This, for what it's worth, was an idea that never seemed to die among oilmen. I said feeding everyone was a very real problem, but for entirely different reasons: it was thanks to oil that the dependable weather patterns needed to grow those crops were breaking down, making crop failure a real risk in many parts of the world. With or without oil, we were going to have to confront the issue of food security eventually.

Unfortunately, I said, the world's current trajectory had been decided by a relatively small group of old men, most of whom were now dead. So certain were these oilmen of their convictions, and so in thrall to their own greed, that they had driven humanity down a pathway to a warmer world. There was no way to regain goodwill after concealing research about the implications of climate change and a couple of decades spent bullshitting about the science of global warming, just as the future was not a matter of 'us' or 'them'. Everyone alive today would suffer the consequences.

'We are all living in their world now,' I said.

The next morning, the oil and gas execs nursed their hangovers as they filed back through the security checkpoint for the final day of presentations. If the mood had recovered into one of casual nonchalance the evening before, today it would deteriorate into something dystopian.

Opposition leader Peter Dutton set the tone with his address. His face was projected onto a 50-foot-high screen in the main theatre as he delivered an angry, rambling rant to stir up an industrial rebellion against the government. Ronald Reagan – the US president who stripped the solar panels from the White House roof – had managed a previous oil crisis through 'deregulation', Dutton said. Reagan did this, he claimed, by letting 'freedom solve the problem through the magic of the marketplace'.

I laughed out loud at the word 'magic' – at this point, this was the best they had.

'Thoughts and prayers,' I said, quietly.

The opposition leader then urged the nation's oil and gas producers to attack the Albanese government over its 'interference' in the gas market.

'The Coalition will fight for your sector with energy, enthusiasm. But we also need you to fight for yourselves. We need you to speak up, frankly, and more avidly,' he said. 'If you don't speak up now, I think it's just going to put the sector at more risk. It's our country's future prosperity we're talking about. And we need you to work with us to push back against this government's detrimental policies.

'We need you to work together to promote nuance and inject reason into the public debate. We need you to lobby your local representatives, and to do all these things with solidarity in your sector. I think now is the time to seize the opportunity to speak up.'

Reason. Freedom. Solidarity. It was strange to hear the conservative opposition leader talk like Lenin. But there we were.

Outside the conference, things were escalating. Earlier that morning, a pair of Extinction Rebellion activists had splattered paint on the ground-floor glass windows of the Santos office building. Unfortunately for the activists, they had missed their target. The lobby of the building was occupied by an independently operated café, and the owner immediately took to social media to complain about being caught in the crossfire.

The image of a small business owner in distress curdled the outrage brewing on talkback radio overnight and South Australian Liberal leader David Speirs sensed opportunity. Calling into Adelaide's talkback station, FIVEaa, he made the all-too-predictable call for a crackdown. 'These types of protests are getting out of hand,' he declared, 'and we are sick and tired of seeing groups and individuals receive nothing more than a slap on the wrist.' He wanted to ratchet up the fine for disruptive protests from $750 to $50,000 or a three-month jail sentence. This fine was double what had been imposed in New South Wales – laws which the New South Wales Supreme Court would overturn in December 2023.

Follow-up callers applauded this opportunism, some half-joking about whether it might be possible to bring over 'South American-style water cannons' to disperse future rallies. Any other time this sabre-rattling might have ended as limp words in dead air, but within half an hour, South Australian premier Peter Malinauskas intervened, calling in to the same radio station to hammer out the details. Having apparently turned a talkback radio studio into a third house of parliament, the premier said his government would be more than happy to back the changes but needed 'bipartisan support' to get them done. Were he to get this unity ticket, he reckoned they could probably deliver it same day.

It was a standard play for the Labor premier. When his opponents caught on to an idea he felt had popular support, his instinct was to assimilate it and present it as his own to stop his rivals from getting oxygen. Normally, the stakes weren't high enough for anyone to mind these 'captain's calls', but what was being discussed was a fundamental challenge to a basic civil liberty – the right to protest. Many Labor parliamentarians had cut their teeth protesting one issue or another, and the unions that formed the party base were regularly engaged in disruptive rallies. What the premier saw as a reasonable response to the threat seemed more like a long-term constraint to others within his own party. It was all well and good to give the courts greater capacity to inflict punishment under an enlightened Labor government – but were a union-busting Liberal government ever to take power, things might be different. The more militant unions would bear the brunt long after Extinction Rebellion was gone.

Worse, the premier seemed entirely oblivious to how it all looked. Tom Koutsantonis – who had no love for Extinction Rebellion after its members had embarrassed him at another petroleum conference earlier in the year – had told the industry that the South Australian government was 'at your disposal' just two days before. On top of that, the premier's brother, formerly Koutsantonis' chief of staff, worked for Santos, having been offered the job a month before the state election. When reporters learned about this a week after Malinauskas was elected, the premier swore he would 'make all appropriate declarations' when dealing with his brother in an official capacity. His brother likewise said he would remove himself from any direct dealings with the premier. Santos, meanwhile, told reporters that any suggestion of a conflict of interest was 'disrespectful'.

There was no evidence of any collusion in the push for the anti-protest laws, and the premier would later reject any suggestion that his personal relationships had any bearing on his decision-making. However, the existence of this and other relationships was enough to fuel speculation – if not outright conspiracy theories. It was, as those in public relations say, a bad look.

The first I heard about the new anti-protest laws came from a startled text message that coincided with an appearance by Santos CEO Kevin Gallagher on a panel I was watching. Gallagher was explaining how the oil and gas industry wasn't about to go to war, but now that they were 'in the firing line' they had to 'fight back'. The text message had a link to a news report and asked: 'You watching this?'

I was not. As I caught up, I learned that the government had flown into action straight after the premier called into the talkback radio station with promises to rush through legislative change. Labor's MPs were in a closed-door emergency meeting where no-one could reach them. Considering the pace set on most issues before parliament, the speed with which the government was moving was staggering.

As I walked over to the next session, I called the premier's media rep to find out what was going on. I had so many questions: Where was this bill? What exactly did it say? Who had the premier consulted beforehand? What had *they* said? Had he even let the crossbench know?

No-one knew anything. The best the media rep could do was say they would get back to me.

Later, a rumour circulated about an internal revolt at that meeting. Labor MPs had threatened to cross the floor but backed down under threat of expulsion from the party – though no-one would actually confirm this on record. When Labor finally broke cover, the media rep called back to say they didn't know where in the world I was exactly but there was a snap press conference being held over in parliament in 15 minutes.

As I left the Convention Centre for the conference, I pulled up a live feed on my phone and watched the lower house start the process to wave through the bill. Within a matter of hours, a profound limitation on the right to protest had moved from the airwaves to parliament, and no-one other than the state's attorney-general, Kyam Maher, had actually seen what the bill said.

Up in the antechamber of Old Parliament House, we, the assembled press, chattered among ourselves and worked out the key questions about the proposed change as we waited for the premier. It was possible, we thought, that the entire bill could be waved through in less than twelve hours if the upper house played along. The words 'extraordinary' and 'remarkable' were used multiple times.

Walking out to address the firing line of television cameras, the premier pitched his government's response as very sensible. It was his duty, he explained, to ensure the parliament reacted swiftly to the protests sweeping the city. Whether he intended it or not, though, the bill had a beneficiary – the nation's oil and gas industry, who were all hanging out just down the road. In an effort to pre-empt criticism, the premier pointed to South Australia's decarbonisation record, saying, 'The actions of Extinction Rebellion, quite frankly, do the cause of decarbonisation harm.'

Taking his turn, Attorney-General Kyam Maher attempted to frame the change by presenting the activists as aggressors. He claimed the government simply could not stand by while 'other people's freedoms are being impinged' by disruptive protests and that 'lives were potentially put at risk by such protests'.

'If you are doing things that could potentially stop, for example, an ambulance getting to a hospital, then the court has a greater range of penalties they could impose on you,' Maher said.

And there it was. The same mythical ambulance that had screamed its way through New South Wales Parliament had turned up in South Australia to flash its lights. There was no more powerful a rhetorical tool to justify a knee-jerk response to a moral panic, and no better symbol of the moment. It was the perfect metaphor, evoking the terror of helplessness, that feeling of having your hopes, dreams and fears packed into the back of an emergency services vehicle, slowly bleeding out on the floor.

Having made their case, next came questions.

'Is that a hypothetical?' I asked.

'For example, if you block off a road that has a hospital on it, you are unnecessarily causing the potential to block an ambulance. That just stands to reason,' Maher said.

Did it? I thought. I had watched the protest at the bridge take place. There had been plenty of access along the tramway for an ambulance to make it through. But there hadn't been an ambulance, and Maher had no proof to justify such a claim. Although in fairness, with the pace the government was moving, it hadn't had much time to find any.

I tried again. 'So, there was no ambulance?'

'If you're blocking a road, but it has a hospital where an ambulance has to go through, that is a potential consequence of your actions. That just stands to reason,' the attorney-general said.

In neither Sydney nor Adelaide had an ambulance been blocked. Extinction Rebellion made it a point to account for the possibility of emergency vehicles in their protests. Maher simply wasn't willing to admit it in front of the television cameras.

When I tried asking the premier about Koutsantonis' earlier speech, he didn't seem to know about it. Instead, he gave a long monologue about the need for hydrogen to replace gas, saying that he and Koutsantonis had 'a very firm view that gas has an important role to play as a transition fuel, but the sooner we can get off gas, the better.'

It wasn't what I had asked, but by then it almost didn't matter.

After the press conference, the Labor party machinery locked in behind its leader, defending the decision by urging those concerned to read the proposed laws. The catch was that the bill hadn't even been made public at the time it was announced. When it finally was, what it actually said was worse than anticipated. It didn't just ratchet up the existing penalty, allowing fines twice as large as those introduced in New South Wales, it made it possible to jail people who 'directly or indirectly obstructed the free passage of a public place'. What this actually meant was not clear – nor was who could decide whether a crime had been committed. It seemed like it came down to whatever a police officer felt in the moment. The language was so broad that officers at the scene could selectively enforce the regulation depending on whether or not they agreed with the protestors' cause.

And it went further. The South Australian police commissioner, or 'relevant entity', was given the right to bill protesters for 'reasonable costs and expenses'. As written, this meant the police commissioner could charge whatever they determined the cost to have been – unless the offender could prove otherwise, which they couldn't, because that would require access to police financial records.

Uproar at the Labor caucus meeting that night would force the premier to hold off on passing the bill for two weeks, allowing a broad alliance of civil society groups and the union movement to assemble in order to fight the changes. This challenge would have no effect. Faced with a choice between backing down and damaging his brand or angering his own party, the premier chose the second, on the basis that the voting public would forget. By the end of May, the changes would pass into law, with three amendments watering down the worst elements of the bill after a 14-hour filibuster by the crossbench that ensured the process was as painful as possible. When it was done, the premier celebrated the win as 'a victory for common sense'.

Following the presser, I walked back over to the conference to file my story on the anti-protest laws and spent the rest of the day loitering in the lobby. Across the way, Extinction Rebellion had set up speakers and a turntable in the central plaza at the convention centre. Blasting music at full volume, they were now holding a dance party, ringed by police officers who silently looked on.

An older woman in a plain grey, two-piece skirt suit stood staring out through the plate-glass windows at the protesters gathered in the square. Her raised hand fluttered at her mouth as if she were in deep shock. It was a pose that drew attention to the pearls that ringed her neck and studded her ears.

She wore a look of anguish. Tut-tutting the protesters, she caught my eye and gestured out through the window at the scene.

'Do you see this?' she asked, as if looking for validation. 'Isn't this outrageous? All these police resources tied up just for a dance party.'

I didn't know what to tell her. Last I checked, we still lived in a democracy. The central plaza was a public place, and the crowd of punks, ageing hippies and rabble-rousers had just as much right to it as respectable titans of industry standing behind a line of cops. Besides, it looked a lot more fun.

If I had set out in the course of my reporting to show that Australia's oil and gas companies weren't 12-foot-tall giants, it seemed a fitting end. But then, as I had learned, neither were they mice. Australia might have been late to the party with oil exploration, but the industry's history had played out much as it had in other places. The US and British firms that underwrote its beginning knew their actions were contributing to a problem that, with enough time, would present a threat to humanity – and they'd done it anyway. At first, they might have been willing to talk openly about climate change as men of science and industry, but that had faded. It was simply too hard to swallow the idea that the very thing they excelled at, the work they had devoted their lives to – the same work that made them filthy rich – might destroy civilisation.

The world needed them – they insisted upon it. They believed they were heroes. Part of the solution. And they weren't going to go quietly. They would stand up, they would tell their story, and people would hear their voice. If people didn't hear them, they would just have to speak louder.

I stuck around for the chest-thumping closing speeches in which they said all this, then skipped the post-conference cocktails to join the crowd making straight for the exits and left quietly. I'd seen enough. Along the way, a couple of protesters stood at the railing, politely wishing attendees a happy conference but holding signs reminding them they were cooking the planet.

Heading down the escalator for the last time, I saw it. At the bottom, outside on the pavement, someone had written a message in chalk for the suits:

'You have a choice.'

16

A Tipping Point with Irreversible Consequences

We were outside, standing in the patio section of a French restaurant directly across from the Australian Pavilion. It was somewhere around 7 pm in Dubai, and this was the closest place that sold beer, making it a natural fit for the Australian delegation drinks at the 28th Conference of the Parties (COP28).

The event was a fixture at these conferences, and the attendees weren't just limited to Australian ministerial and diplomatic staff. Anyone with even the loosest connection to Australia found a way to attend to take advantage of the free booze. Among them were a noble collection of scientists, researchers, policy wonks and a broad array of civil society groups who had come to watch negotiating teams hammer out a global response to the existential threat of climate change on behalf of almost 200 governments. Behind the ministerial and departmental staffers travelled the usual gaggle of camp followers – smiling CEOs who dispensed cards when you shook their hand, washed-up politicians grasping at relevance and lobbyists desperate to get a photo with the minister for their socials.

My friend slung me a beer as we surveyed the scene. He was an economist with more bachelor's degrees than sense and a wealth of experience working as far afield as Washington and Papua New Guinea. Now he was in environmental policy with a civil society group, and, like me, this was his first time attending the international climate summit.

We had both come to bear witness to what felt like a historical moment. This COP was supposed to be pivotal, the first to directly confront the future of oil, gas and coal production – but after a couple of days on the ground, any open-minded sense of possibility had curdled into cynicism.

'None of this is about the environment for most of these people,' my friend said, surveying the crowd. 'Nature is not even a secondary concern.'

To remove ourselves, we made our way to an empty table at the back of the patio section. We had just sat down when a strange man approached, asking for a seat at our table. He was a short, portly fellow who spoke with a smoothened Northern Californian accent and seemed friendly enough. We obliged and he collapsed into his seat, wiping the sweat from his brow. Steadying himself, he leaned forward, peering over the table at us through his thin, wire-frame glasses to get a better look at our conference badges. Spotting that I was a reporter, the man asked me a question.

'You're not going to quote me, are you?' he said. 'You're an honourable man now, aren't you?'

I should have found it odd, but I was too tired to notice. Squeezing into a packed, driverless metro train that charged through the desert haze, I'd arrive early each morning and make my way through a kilometre-long cattle run to security. All the while, the taste of desert dust and pollution settled on my tongue. I'd spend the next ten hours trekking a vast concrete expanse between press conferences, interviews and pavilions under the bright Arabian sun. After a few days of this, my feet hurt, my patience was thin, and I had no idea who this man was. I also didn't particularly care.

'I'm just here for the free beer,' I said.

The man at our table asked how we felt about events so far at COP28. In my view, I said, there was only one issue that mattered now: fossil fuel phase-out. Of all the things people were talking about on the sidelines of the negotiations, the most important issue was about the future of the oil and gas industry.

I'd spent the last few days watching impassioned presentations about the need for action. I'd also watched grifters give presentations on subjects like sustainable yachting, green bitcoin mining, artificial intelligence as a solution to climate change, and so-called e-methane – an allegedly carbon-neutral fossil fuel. A couple of them had even given me their cards. Arguably the two most powerful men in the world, US president Joe Biden and Chinese president Xi Jinping had given COP28 a miss, but Russian president Vladimir Putin trolled the gathering when he turned up next door in Abu Dhabi to negotiate future oil exports with the United Arab Emirates.

All the big oil executives, however, made sure to put in an appearance. ExxonMobil CEO Darren Woods was one of several executives from the world's biggest oil companies to land in Dubai, the first time these climate villains had ever personally attended the international climate summit.

And they couldn't have chosen a better moment. It had just been declared the hottest November on record globally, and over 100 countries had grouped together to call for a 'fossil fuel phase-out' at COP28, demanding that governments agree to a clear timeline for a managed end to oil, gas and coal production. For three decades these fossil fuel producers and their industry associations had operated from the shadows to keep an eye on the international climate negotiations that took place with each COP. Their open presence at COP28 felt like they had simply stopped pretending to be disinterested.

It helped that the guy overseeing the whole show was one of their own. Dr Sultan Ahmed Al Jaber was a tall Emirati technocrat who spoke in a deep, resonant rumble and always seemed to appear in public wearing a traditional white kandura. From the moment his appointment as COP28 president was confirmed, there was controversy: Al Jaber's day job was CEO of the Abu Dhabi National Oil Company (ADNOC).

To be clear, Al Jaber wasn't just an oilman: he was a creature of the oil company. ADNOC had sent him to school as a young man and given him his first job. Over the course of his career, his public image as a new-generation Emirati technocrat had been carefully sculpted by Edelman, another significant public relations firm relied on by the US oil and gas industry and the same firm Al Jaber had contracted to help organise COP.

No less than former US presidential candidate John Kerry and former UK prime minister Tony Blair lobbied aggressively to place Al Jaber as COP28 president, citing his previous work in renewable energy development as a qualification. When critics protested that letting an oil executive oversee the international process that was supposed to fix climate change was like putting the fox in charge of the henhouse, his supporters argued that it was a political masterstroke and a cunning trap. As an oilman, Al Jaber could 'bring along the oil producers' to effective action on climate change, they said, and the intense scrutiny he'd be under, and of the United Arab Emirates more broadly, would ensure good behaviour. They were presenting Sultan Al Jaber as the oilman who would end the global oil industry.

It was a naïve fantasy that ignored how, under Al Jaber's leadership, ADNOC was planning to expand production at the state-owned oil company, not slow it down. Neither Al Jaber, nor the sheikhs he reported to, had any intention of winding up the industry. In the opening days of COP28, an explosive investigative report by the BBC and the Centre for Climate Reporting revealed that ADNOC personnel had been intimately involved in preparations for the conference, despite promising to keep the oil company separate, and that there were plans to use Al Jaber's access to government delegations to strike new oil deals. Confronted with these allegations, Al Jaber denied the reports, claiming that the western media were simply out to get him. Some individuals associated with the United Nations Framework Convention on Climate Change (UNFCCC) privately told me that they agreed with him. The media, these people said, were too interested in black-and-white narratives of 'good versus evil'.

This was not a universal position – others thought the press should be going harder – but it was enough that I began to wonder whether those tasked with advising and administering this process – all clever, highly qualified scientists in their own right – lacked street smarts. It wasn't just the COP hierarchy that people felt had been compromised by associations with fossil fuel producers. Among the representatives of civil society groups on the official attendance list, at least 2400 had ties to fossil fuel companies – and at least 160 were known climate deniers. In total, delegates with oil, gas and coal connections at COP28 outnumbered delegates from the Pacific Islands by a factor of 12. A representative of BP was also counted among Australia's 'party overflow' delegation – that group of people who didn't work for Australia in any official capacity but who still had enough clout to secure an invite from the Australian government.

And for the first time, the oil industry had even been given a physical presence at a COP.

Out of curiosity, I had wandered over to the Organization of the Petroleum Exporting Countries (OPEC) booth, tucked away at the back of the conference site in a building marked 'Urbanisation & Indigenous Peoples'. A well-dressed man spotted my press badge and did everything he could to ignore me as he attempted to sell a young hijabi on a career with the oil cartel. When I tried to ask him how it was going, I was told OPEC secretary general Haitham Al Ghais would answer all my questions. As I turned to leave, the man seemed relieved.

In a manner of speaking, he was correct. By the second week, the bloc of countries calling for a commitment to a fossil fuel phase-out was large enough that there was a growing expectation it might actually happen. The situation apparently spooked Al Ghais so badly he wrote to OPEC's member countries urging them to block any commitment to a fossil fuel phase-out. The letters immediately leaked.

'It seems that the undue and disproportionate pressure against fossil fuels may reach a tipping point with irreversible consequences, as the draft decision still contains options on fossil fuels phase-out,' Al Ghais had written. 'It would be unacceptable that politically motivated campaigns put our people's prosperity and future at risk.'

As a group of oil-producing nations, OPEC was just one bloc of fossil-fuel dependent economies, operating mostly in the developing world. Another bloc included developed economies like Australia,

Canada, Norway, the UK and the US. Publicly, these countries had all signalled their commitment to a phase-out, but observers frequently countered that their words did not match their actions. Australia, alone, was considering at least three dozen new proposals for oil, gas and coal projects.

If negotiations failed to deliver a phase-out, I told the unidentified stranger at my table in the French restaurant, the whole COP process faced a crisis of legitimacy. Thirty years since the United Nations Framework Convention on Climate Change was negotiated, the bill had come due. As I had spoken to people around the conference, many had been frank in asking: 'What is the point of COP?' I suggested those calls would only grow louder if action didn't follow.

The man at our table waved off these concerns. He had been coming to COP for the last 25 years, he said, and it was always the same. These international climate summits were filled with stories of victory snatched from the jaws of defeat. Success, he said, depended on some special mix of trust, charisma and the personal organisational ability of the COP president.

It was at that point I finally asked the man who he was and what he was doing there.

'That's why I asked you not to quote me,' he said.

I sat and listened as he explained that he worked for an oil company: a direct descendant of Standard Oil and one of the biggest in the world. He was officially there as an observer, wearing an IPIECA badge. His job wasn't to lobby, he quickly added, but to report back to the company on what transpired in Dubai. He was a veteran of these processes, and a trained scientist who had even contributed to a report produced by the IPCC. These IPCC reports were highly complex works that distilled the best-available knowledge of science at the time. In many ways, they were a pinnacle of human achievement – so much so that, in 2007, a Nobel Prize had been given to the collective authors of the Fourth Assessment Report. As a contributor to the section on carbon capture and storage, the man at our table was very proud to have been included.

I listened in quiet disbelief. If I had travelled around the world to find out how far the influence of fossil fuel companies extended,

I had found it. The Conference of the Parties was supposed to be the international mechanism by which humanity dealt with an existential threat to its existence, but the oil, gas and coal producers had correctly identified it as an existential threat to their business. Over time, they had sought to shape it. This man wasn't solely responsible; there were hundreds of others like him running around. But as the physical manifestation of the phenomenon, he had sat right down at our table.

Out of curiosity, I asked whether he knew anything about IPIECA's role in coordinating the campaign to undermine the science of climate change.

'That was before my time,' he said. 'Like you, I'm just here for the free beer.'

Late on the following Monday, the draft of what was supposed to be the oil, gas and coal industry's death certificate finally dropped. The whole point of COP28 was to decide the final language of the first Global Stocktake under the Paris Climate Agreement. This was an exercise in working out how bad the problem was getting, followed by an effort to identify what action governments needed to take to address it.

At this stage in proceedings, delegates from the 199 parties to the UNFCCC had spent the last two weeks in various committee meetings and plenary sessions hammering out the text of this document. The stakes had been high in the hours leading up to its release, as were expectations. There had been several drafts released since COP28 started and pressure had been building. A few days before, UN secretary-general António Guterres had flown in to personally urge delegates to 'end the fossil fuel age'. Meanwhile, COP28 was scheduled to finish on Tuesday, and the Emirati hosts had made plain their intention to finish by deadline. With the clock ticking down, the expectation among civil society groups and observers was that a commitment to a fossil fuel phase-out was coming, perhaps with some 'artful' language to avoid obstruction by petrostates like Saudi Arabia and Russia.

When it came, the draft text did nothing of the sort. In the critical section that was supposed to deal with the transition away from fossil fuels, there was no mention of oil or gas, let alone a commitment to a phase-out. Its provisions were riddled with caveats and qualifications. Despite acknowledging the science of climate change and the growing risk of measurable, material harm to lives and livelihoods, the best it offered was a shopping list of actions it suggested countries 'could take'. These included 'reducing both consumption and production of fossil fuels' and a phase-out of 'inefficient' subsidies. It even went so far as to wind back language agreed upon at previous COP summits: an existing commitment to phase down coal production was now a commitment to phase out 'unabated' coal production and 'unabated' coal power generation. As no-one could properly define what 'unabated' meant, it was a convenient loophole for countries looking to open new coalmines or build new coal-fired power plants.

The response was mutiny.

As soon as the text was posted online, it began to percolate through group chats and social media platforms, instantly animating scientists, civil society groups, Indigenous peoples and activists. Phones lit up with furious text messages, and scathing analyses were posted.

The world's press, who were tetchy after having, once again, spent the day waiting for something to happen, began to gather outside the plenary hall in expectation of a press conference with Sultan Al Jaber. The hall was a huge, generic modernist construction of cement, glass and steel that looked like any other airport, university hall or conference building. A roped-off area had been set up with a dais and photo wall. Half an hour passed, and the atmosphere in the public area grew increasingly tense. When word finally came that the press conference had been cancelled, the waiting reporters grew riotous.

'What is he afraid of?' someone yelled.

As this was going on, government delegates, scientists and activists had raced over to the media centre, perhaps an 800 metre walk from the plenary hall, to provide their reaction. The reporters who heckled Al Jaber piled into a little open-air courtyard outside, no bigger than a hotel reception area. Cameras flashed and boom mics rose above the tightly packed cluster of bodies. Outstretched hands holding mobile phones ringed the talking heads as, one by one, they denounced the draft text as a failure, a capitulation to fossil fuel producers, a death warrant for Pacific Island nations – and for humanity.

At this point in human history, every serious person recognised the danger. Global food systems depended on regular, dependable weather patterns. Cities had been built and optimized to function within a particular temperature band. As the world's climate systems were breaking down, so too was the environmental niche which we inhabited as a species. This was not some mysterious act of god, or some cruel act of a random universe lashing the planet with an uncontrollable cosmic force. The cause was simple, mundane and knowable. Burning oil, gas and coal was changing the chemical composition of the atmosphere – an act of slow violence inflicted upon the majority of the world's population by a relative few. This violence was not immediate and obvious like a mugging, but more gradual, so as to be nearly imperceptible – and arguably more lethal. It took generations of co-conspirators to carry out and was now delivering a growing body count. The harm was also likely to continue long after the problem had been solved owing to the lag times in complex, global climate systems. Those who recognised the problem and wanted to do something about this had already been forced to waste so much time confronting attempts to sabotage action. Humanity had already spent a decade debating whether it was actually happening and then another trying to satisfy ever-increasing demands for evidence from bullies and bad-faith actors who could never be satisfied.

What they wanted was for COP28 in Dubai to be the moment where something changed – a *deus ex machina* of the human spirit. All those people standing in the courtyard outside the media centre had come to demand that those representatives acting on behalf of all humanity in 2023 do the obvious and inevitable – to finally agree on a pathway forward for meaningful action. For their calls to be answered with a collective shrug and apparent indifference to the emerging catastrophe that everyone acknowledged was occurring, was cruel mockery by the rich and powerful whose wealth, status and prestige depended on the carbon-economy.

With the initial outpouring of rage and frustration now on the record, the reporters peeled off to file their stories and tell the world how we had all been sold out. Everyone else went away to grapple with the prospect that everything they had been working on, all their efforts on climate change to date, had been pointless – they were witnessing the betrayal take place in real time.

It is impossible to know what went on behind the scenes, but judging by what came next it was clear a thousand different angry conversations had begun, leaving Al Jaber rattled and prompting fresh negotiations about how it should be rendered.

As the hours wore on, unease grew. It soon became clear these deliberations were not going well and that the hosts' deadline would not be kept as COP28 ran into overtime. There would be no press conferences or background briefings over the next 48 hours – though at one point close to the end, the staff of Australia's climate change minister Chris Bowen summoned select reporters for an off-the-record chat – I was not among them. The only indication of how things were going came from a few loose-lipped British diplomats. Speaking loudly while in line for coffee in the UK pavilion, they were quite open in saying negotiations were two days behind where they ought to be.

Around this time, the conference venue began to empty. Representatives from NGO's began to fly home, and government delegations started dismantling their pavilions. Before long, the only people left were roving bands of reporters waiting around for some kind of resolution before they too could ship out. Many of them retired to a hotel bar around the corner from the media centre as Tuesday bled into Wednesday without any end in sight. Then, around 8 am the next morning, everything changed.

★

Booming applause welcomed Sultan Al Jaber as he took his position at the microphone to deliver his closing address. Al Jaber was dressed in his pristine white Kandura and as the camera settled on him, it framed a perfect midshot. A broad smile swept his face and his eyes bulged as he basked in the world's approval.

COP meetings were infamous for their drama. The last plenary session was supposed to be given over to approving the final text of any global agreement outlining what the parties should be doing to address climate change. In practice, it often dragged out, as procedural objections, rebellions and motions forced eleventh-hour tweaks.

Not this time. In Dubai, the production and presentation of this document to the world would be carefully stage-managed. Al Jaber had been meticulous in ensuring that, on his watch, the world would not be presented with a display of disunity and division. His strategy had been to privately secure agreement on the final text with all parties before the television cameras were switched on, ensuring the only thing the world saw was a well-managed spectacle.

When the oilman introduced and then gavelled through the final text from COP28 – which he dubbed the 'UAE Consensus' – without objection or opposition it took everyone by surprise. It had all happened so quickly. Within minutes of the opening plenary starting, it was over: the world hadn't agreed to kill the oil, gas and coal industry. The rest was all detail and ego stroking.

At best, the text of the Dubai agreement amounted to 'the beginning of the end' of fossil fuel production. The agreement did many things, some of them even good, but the key section – clause 28 – did not bind any nation to the phasing out fossil fuel production. Instead, it 'calls on' governments to '[transition] away from fossil fuels in energy systems, in a just, orderly and equitable manner, accelerating action in this critical decade, so as to achieve net zero by 2050 in keeping with the science'. It included what Al Jaber described as an 'action plan', with several options for countries to pursue lower emissions. According to the COP28 president, the 'historic' document marked a 'paradigm shift', as it was the first time the COP process had directly addressed fossil fuels.

Whether it was enough was another question. Even as political leaders worked hard to frame the compromise as an unequivocal win, others pointed to loopholes large enough to pilot a gas tanker through. The text said nothing about fossil fuel subsidies or the opening of new fossil-fired power plants. A new clause was inserted stating that the parties recognised 'that transitional fuels can play a role in facilitating the energy transition while ensuring energy security' – a common fossil fuel industry talking point. And even as it sought to nudge the world away from fossil fuel production, the agreement entirely failed to address an important issue: rich countries had yet to actually stump up the cash the developing world needed to transition away from oil, gas and coal. This issue would be left to the next meeting in Azerbaijan. At best, COP28 had produced a plan to create a plan to deal with a problem everyone had finally agreed to call a problem.

It took Anne Rasmussen, lead negotiator for Samoa, to puncture the air of smug self-satisfaction. She had been outside the plenary hall when the gavel fell, wrangling delegates from other low-lying islands in the Pacific. The Alliance of Small Island States had formed to allow these nations to band together, giving them strength in numbers, and it was they who had repeatedly pushed the hardest for a fossil fuel phase-out. Rasmussen had entered the plenary hall during the standing ovation given to the COP28 president and was initially confused. It was only when she sat down that she realised it had all happened while she was out of the room.

Upon being given permission to address the assembly, Rasmussen explained she and her fellow Pacific Islander delegates could not lend their support to the statement.

'We have come to the conclusion that the course correction that is needed has not been secured,' Rasmussen said. 'We have made incremental advancements over business as usual, when what we really needed is an exponential step change in our actions going forward.'

The agreement failed to deliver the action Pacific countries had been demanding to stop the ocean consuming their homes, she said. It failed to commit world governments to the action that scientists said was needed to constrain global warming to 1.5°C. Outside they had resolved not to oppose the motion, but with it now having passed, the Pacific Island nations sought to register their protest, knowing their wishes had not been respected and that the world had failed them.

'It is not enough for us to reference the science and then make agreements that ignore what the science is telling us we need to do,' Rasmussen said. 'This is not an approach that we should be asked to defend.'

At best, COP28 could be set to have started the clock on the industry's demise, even if it didn't specify when time will be up. If the global oil industry had bought more time in Dubai, back home news came that sent a chill up the spine of the Western Australian and South Australian state governments. Word had leaked that the country's two biggest domestic oil companies, Woodside and Santos, were in early talks about the prospect of a merger. The companies stressed they were just talking and that there was no guarantee these conversations would lead anywhere. But if they did, the combined heft of the resulting $80 billion entity would make it one of the biggest oil companies in the world.

These talks went nowhere and were abandoned in February 2024 but the fact they happened is significant. It might seem counterintuitive but the spectre of two Australian oil companies discussing whether to join forces to create a corporate behemoth was not a sign of strength. Globally, those who ran the world's oil companies were aware that peak demand was approaching as the uptake in renewable energy and electric vehicles increases. So far, only one oil company in the world, Ørsted in Denmark, had successfully embarked upon a process of transformation. Not all would or could make a similar transition out of oil production and into something else. Many executives felt it would be wrong for them even to try. According to the logic of finance, an oil company was in the oil business, not the solar or wind business. If the world was moving away from fossil fuels, the 'appropriate' move was not for the oil company to change 'core business' but to release shareholder capital back onto the market so it could be reinvested elsewhere, in more profitable – and less lethal – operations.

Even as the industry was collectively fighting to continue, individual companies were beginning to quietly make moves in this direction so as not to spook the market. Around the world, the big oil companies were picking off their competitors and ploughing their super profits into share buyback schemes. The point of this consolidation was to generate mass. Over the next decade, those operating alone could very easily be caught out by falling demand and rising production costs. Achieving scale meant these companies had a shot at being the last guys standing.

This was the long game these executives and their corporate ancestors had been playing for half a century. In an industry where long-term strategic planning spans decades, ignorance is no defence. These people learned long ago about the harm caused by releasing CO_2 into the air. They knew the risks in upsetting the chemical balance of the atmosphere, and their research had only deepened their understanding. They had grasped before anyone else how the lag times between input and effect meant the consequences of their actions would still be felt long after they were gone. These men and women happily mortgaged the collective future of humanity in the name of shareholder value and multimillion-dollar bonuses. Having seen the writing on the wall before anyone else, their instinct was to play for time. The longer any reckoning could be drawn out, the more time they had to run their operations into the ground, extracting every last cent, pumping every last drop.

By playing for time, the rich and powerful have so far escaped efforts to hold them accountable. The greater the distance between action and consequence, the easier it is to dismiss questions of accountability as historical – a theoretical debate between stuffy academics, of limited interest and with no material bearing on the present. The challenge is to reject this framing. Some degree of climate change may have been inevitable, but the collective crisis we are now facing was not. The collective future we now share was never inevitable; there was always a choice – and the record shows industry sought to deliberately engineer those choices to produce certain outcomes.

What we need is a clean break from the past; a conscious effort to shake off the legacy of state and institutional capture at a time when the industry's grip on power is weakening – a bare-knuckle attempt to examine the greed and moral apathy that led us here and make the political, legal and social ramifications of the oil and gas industry's choices felt. A managed end to fossil fuel production, litigation and compensation are necessary but should go further. Industry can no longer be afforded a social licence – they have burned their seat at the table. As they have shown they cannot be trusted, they cannot be allowed to shape public policy decisions. Individuals can refuse to cooperate and organisations can stop taking their money.

Meanwhile, across the country, industry archives remain locked, preventing a full accounting of events. These should be cracked wide open and made public, with access managed by libraries or universities. Until then, those with access should leak everything they can. Accountability requires evidence and so long as these documents remain buried in dusty collections, largely forgotten, it is impossible to make a full and frank accounting of what occurred, especially as the industry, so far, would prefer not to reflect too hard on its past.

For now, though, the oil and gas industry refuses to die. And if they are pushed, they are determined to ensure it is they – not government, not the people – who dictate the terms of their exit. With twilight closing in, you can be sure the men and women of the oil industry are busy planning for a very, very comfortable retirement.

The only question is whether we will let them get away with this, too.

Acknowledgements

Writing a book is not something you do alone. A book takes many hands to make, and this one is no different.

None of this would have been possible without financial support from the South Australian government through the 2023 Arts South Australia Fellowships program. My fellowship allowed me to travel to conduct fieldwork, pick through archives, interview people and gain firsthand, on-the-ground accounts. No other additional external financial support or contribution was received in connection with the project beyond what I earned as a freelancer.

I owe a deep debt of gratitude to several people who provided expertise and technical guidance on aspects of this project in one way or another. They include Dr Grace Augustine, Tim Baxter, Dr John Cook, Kert Davies, Kate Dilger, Ian Dunlop, Neil Francey, Roger Gifford, Dr Marc Hudson, Kirsty Howey, Ketan Joshi, Carroll Muffet, Belinda Noble, Jesse Noakes, Dr Lily O'Neill, Dr Ellie Piggott, Lyndal Rowlands, Dr Birthe Soppe, Dr Russell Tytler, Dr Jeremy Walker and Marian Wilkinson, as well as a host of others who cannot be named. There are several others who provided assistance during research, including Nicola Badran, Walter Marsh, Sienna Parrott, and Lexie Seager.

A special thank you to Lyndal Rowlands and Dr Marc Hudson,

who kindly read early draft versions of the full manuscript prior to publication to offer their thoughts, insight, technical knowledge and experience.

I'd like to acknowledge the time and work of photographer Isabella Moore, who accompanied me on three trips and contributed her photos to this project, and the editors I worked with on the individual pieces of journalism that formed parts of this book, or the manuscript as a whole: Patrick Keneally, Adam Morton, Poppy Reid, Erik Jensen, Emily Barrett, as well as Elizabeth Cowell, who edited the manuscript. All helped save me from myself in one way or another. I'd also like to acknowledge Julian Morgans, who assigned me my first stories on climate change for *VICE*, which prompted me to begin thinking about these issues.

I would like to place on record my deepest thanks and an apology to my agent, Melanie Ostell, whom I forced to wait years while I completed this project. The same is true of Aviva Tuffield, my publisher at UQP, who gave me the opportunity to chase this story. Thank you to Jacqueline Blanchard and the entire production team at UQP, who have worked with me to edit, proof and print these pages. Thank you all for helping bring this thing into existence.

A special thanks must go to Jessica Alice, my partner, first reader and biggest supporter, who tolerated me throughout all the months I spent writing this manuscript with considerable grace. This book would not have been possible without your clear-eyed insight and editorial instinct – the title was her suggestion.

I would like to say thank you to my colleagues from *The Guardian*, *The Sydney Morning Herald*, *News.com*, *The Australian*, *The Financial Review*, *InDaily*, *The Advertiser*, *Rolling Stone* and the BBC, whose work has informed aspects of this story and helped me to reconstruct events.

Thank you to all those named in this book, and those who could not be named. And thank you to the long list of people who shared names, numbers, insights and observations. That list is too long to include here, and some names cannot be included for obvious reasons, but I am grateful to you all.

Thank you to the staff at the National Library and National

Archives of Australia, the State Library of South Australia, the State Library of New South Wales, the State Library of Victoria, the British Library and the Center for American History at the University of Austin. The work these institutions do to maintain the historical record is critical – though I am still waiting for those National Archive requests to be looked at.

Thank you to both the activists who trusted me with their stories and those in industry who trusted me enough to speak to me. Thank you, even, to the media team at APPEA/AEP who – apart from that one guy – handled my requests with professionalism. I know it could not have been easy.

And finally, I'd like to apologise to the long list of friends and family I have abandoned while working on this book, and thank those who offered support, encouragement and distractions. Some of those are: Walter Marsh, Sia Duff, Dr Gemma Beale, Paris Dean, Dr Sam Whiting, Sarah Brady, Alison Whittaker, Tanya Jane Brain, Laurie-May, Dan Thorpe, Dr Michael Hilditch, Issy Bailey, Stephanie Hastie, Matthew Doran, Russell Jones, Max Cooper, Elle Hoffler, Meg Riley, Geoff Goodfellow, Tom Thatch and Leo Corpuz.

Notes on Sources

An industry is composed of hundreds, if not thousands of people, each with their own individual stories. Their everyday decisions and activities shape its operation – and its legacy. The personalities I focus on in this book are a fraction of those who participated in the development of the Australian petroleum industry in one form or another. My purpose in this project was not to give a forensic accounting of events, issues or people, but rather to focus on the 'how'. Among the key questions were how Australia's oil industry started, how it learned to wield influence, how it operated in relation to its international counterparts, and how it used its clout to shape the national response to climate change. In asking these questions, my goal was to scrutinise an industry which is often considered in the micro, but not the macro.

During my research, I asked APPEA whether I could review its archives but was told 'APPEA does not provide public access to its historical materials.' I made a decision early on to quote primary documents at length or in full, where appropriate, rather than breaking up this material and rendering it accordingly. Though it sometimes made the material more difficult to take in, and constrained how I could approach certain stories, my preference was to let the documents speak for themselves. I made considerable effort to focus specifically on oil and gas producers, and not to generalise by exploring coal mining or resource extraction more broadly. The terms 'oil and gas' and 'petroleum' are used throughout the book to refer to both upstream and downstream sides of the business, to keep things simple for the reader.

The observation that, in Australia, climate change is understood through the prism of its fossil fuel experts was first made by Clive Hamilton. I have made a conscious effort not to 'reinvent the wheel' by rehashing history that has been diligently recorded and reported by others such as Hamilton, Pearse and Wilkinson. This is flagged at points within the book; for more detailed accounts of events, I strongly encourage reading Marian Wilkinson's *The Carbon Club*, Guy Pearse's *High and Dry* and Jeremy Leggett's books recounting what took place during the early international climate negotiations.

The part titles that break the story up into Early, Middle, High and Late periods are borrowed from medieval history. This is as much an exercise in black humour as critique. Each chapter title was chosen from among quotes given in that chapter, a technique taken from journalist Geoff Dembicki.

The names of people I interviewed and whose accounts of events I have directly relied upon are given in the text. Only one name has been changed. A list of primary documents relied on in each chapter are collected below, grouped by chapter. They are rendered in accordance with the fourth edition of the *Australian Guide to Legal Citation*, with pinpoint references where appropriate. All conversations were either recorded electronically and then transcribed with Otter AI, or in notes taken contemporaneously. Follow-up contacts were made to fact-check details and quotes.

Some of those I spoke to working within or associated with the petroleum sector were reluctant to speak on record. I relied on information they provided to build an overall account of events, but I have not quoted them directly. The CO_2 figures quoted throughout the book are drawn from Our World In Data, a scientific online publication with a research team based at Oxford University. The contemporary events described in this book took place between the

APPEA 2022 conference and the APPEA 2023 conference. These conferences bookend the work.

I made multiple attempts to contact Keith Orchison, but a message sent over social media plaform LinkedIn was met with a message that read: 'Keith Orchison declined to continue this conversation for now.' An attempt was made to contact Dick Wells through the only publicly available channels but no response was received. APPEA, the Australian Gas Association, Santos and the Australian Industry Greenhouse Network were contacted. I made repeated attempts to contact Michael Holmstrom to further clarify matters we had discussed but received no response. I made a concerted effort to find Elaine Grossman to discuss her work but could not find a current record of her, and her former colleagues did not know how to contact her. Narrabri Mayor Ron Campbell was contacted but did not respond by email. The Australian Workers' Union was given an opportunity to respond to statements made by Misha Zelinsky but 'politely declined'.

The introduction is a reworked version of a story originally published by *Rolling Stone Australia* titled 'The oil and gas industry refuses to die'. In the introduction, I provide the World Bank definition of state capture, a summary of the IPCC's work on climate change and my own observations about the mechanics of truth. This 'model' of truth is based entirely on observations and is intended to make clear how I currently think about the issue; it is not an authoritative statement drawing on academic or legal literature and is not intended to be.

In the words of US journalist Amy Westervelt: every story is now a climate story. In recognition of this fact, chapter 1 is a profile of Kate Stroud, recording how she survived an environmental catastrophe. When climate change is discussed, the focus is often on numbers, political statements and multilateral diplomacy. Kate's story shows what's at stake. I deliberately chose to centre the book on the human cost of climate change, to frame what comes next in a way that makes the implications clear. Through Kate's story, we see how people depend on their environment, how climate change is an issue of generational equity, and also how decisions made in faraway places at other times can have a profound and unforeseeable impact on us, even when we're not paying attention to any of it. The summary of how the public responds to scientists talking about risk is summarised from a Tweet by Lucky Tran.

Kate's story was chosen from among several accounts of a catastrophe in which people died. My goal was to capture the horror, the absurdity and the chaos without lingering on the trauma.

Kate's cautious experience with activism as a source of empowerment is covered in Chapter 2, in which Violet CoCo is also introduced. This chapter illustrates how experience can radicalise people towards activism, while also allowing exploring the legal landscape. The views and opinions expressed by CoCo represent her perspective alone. They are not a representation of my views, or those quoted in this chapter or the wider book. Chapters 2 and 14, both of which feature CoCo, were the product of multiple interviews, field visits and previous reporting for organisations including *The Guardian Australia* and *The Saturday Paper*. I travelled to Lismore for the anniversary of the flood and attended the court hearing in Sydney at which CoCo's jail sentence was overturned and was therefore able to draw on my own direct observations of these events.

Chapter 3 relies on primary documents obtained by the Centre for International Environmental Law that prove the US oil industry had a basic understanding of the greenhouse effect and its implications as early as 1968 and yet did nothing to impede fossil fuel's continued expansion. In my account of the origins of the petroleum industry in Australia, its early political ambitions and its first influence campaigns, I have relied on industry documents and histories commissioned by industry, particularly Rick Wilkinson's *Knights, Knaves and Dragons: 50 years inside APPEA and Australia's oil and gas politics*. Additional context was added through independent research on my part. Information relating to Edward Teller's address at Columbia University and the activities of the American Petroleum Institute is drawn from Dr Benjamin Franta's PhD thesis, 'Big carbon's strategic response to global warming, 1950–2020'.

Chapter 4 charts what is currently known about what the Australian industry knew, and when it knew, using government and industry documents. Dr Marc Hudson identified the 1968 Senate Select Committee on Air Pollution as significant in our conversations, and Dr Jeremy Walker independently sourced and reviewed the evidence given to the committee, identifying Professor Bloom's contribution. Dr Walker shared his annotated copy of this evidence. Notes from the 1971 AMIC conference at which climate change was discussed were found and retrieved by Dr Walker. The sections covering the industry response to the Great Barrier Reef royal commission are drawn from industry documents and Rick Wilkinson's *Knights, Knaves and Dragons*. The activities of Dr Lang were found in a document sweep conducted by Dr Marc Hudson. The section detailing the Australian petroleum industry's earliest known public discussion of the greenhouse effect relies on materials found in the *APEA Journal* archive and the CASANZ archive, with additional context provided by Dr Marc Hudson.

Chapter 5 is a brief historical summary of a turning point in the ideological thinking of the petroleum industry. The sources it relies on are given below, with additional context provided by Dr Jeremy Walker. The speech by APEA founding chair Reg Sprigg was published in *The APEA Journal*. Dr Marc Hudson made me aware of the 1975 meeting in Norwich. Additional biographical detail about Reg Sprigg was taken from Kristin Weidenbach's books, *Rock Star: The story of Reg Sprigg* and *Blue Flames, Black Gold: The story of Santos*.

Chapter 6 tracks how the global oil industry responded to the emerging science of climate change and how Australia responded; it relies on industry histories, peer-reviewed academic work about the role of IPIECA, and Exxon's presentation at the IPIECA meeting in Houston, Texas, in 1984. Much of this material is adapted from Dr Benjamin Franta's PhD thesis. Dr Marc Hudson assisted in retrieving IPIECA documents from the British Library. The final sections rely on industry histories and primary documents, including AIP and APEA annual reports, as well as IPIECA documents.

Chapter 7 retells the story of Noonkanbah, drawing on accounts by Steve Hawke, union accounts and histories published online by the Yungngora. It also draws on correspondence, original reports, press clippings and other materials that record the activities of Len Barker and Joseph Poprzeczny, taken from Len Barker's papers. The historical value of these documents, and the ongoing public interest in the dynamics that shaped Native Title legislation, mean they are described in detail. Though they were late, long and over-budget, the views contained in Barker's reports would shape the industry response to future events and have been credited with 'alerting' the petroleum sector to the emerging land rights issue. I contacted Joseph Poprzeczny while researching this chapter and gave him fair opportunity to respond to a long series of questions. Instead of answering my questions by email, Poprzeczny suggested I travel to Perth to hear his story in person, but I did not have the time or funding necessary to do so. Claims made by Keith Orchison in this period are taken from an industry history and were crosschecked by contacting former Indigenous affairs ministers Fred Chaney and Peter Baume. Orchison did not respond to my attempts to contact him, which impeded efforts to clarify the account.

Chapter 8 builds on my past research and reporting to better understand how petroleum companies have sought to engage with schools as an extension of their public relations program. The details are drawn from industry documents found in the APEA archive, AIP records, annual reports, internal newsletters and commissioned histories collected in public archives. The school materials discussed are available for review in libraries. The section in which I discuss STEM Punks builds on an interview I conducted with Michael Holmstrom as part of a feature report published in *The Guardian*, which is listed below, and is augmented by my observations while attending an APPEA conference and detail drawn from industry documents.

The biographical material in chapter 9 detailing the life and work of Barry Jones was taken from industry documents found in *The APPEA Journal*, industry-commissioned histories and interviews. It was confirmed and supplemented by interviews I conducted myself. I found the AIGN briefing in an edition of the members' newsletter, *APPEA Report*, cited below. The summary of the political history

of the Kyoto Protocol in Australia is drawn from *Scorcher* by Clive
Hamilton, *High and Dry* by Guy Pearse and *The Carbon Club* by
Marian Wilkinson, with input from Dr Marc Hudson. The section
on AIGN draws on Dr Sandy Worden's 1998 masters thesis, 'The
case against carbon tax: the Industry Greenhouse Network's 1994/95
campaign', Dr Patrick Hodder's PhD thesis, 'Climate conflict:
players and tactics in the greenhouse game', and AIGN documents,
supplemented with interviews conducted by myself.

The opening paragraphs of chapter 10 rely on material published
in Guy Pearse's PhD thesis, verified through reference to primary
documents listed below. The background information about
the Lavoisier Group was originally obtained in interviews with
Dr Dominic Kelly for a series of features I worked on for *VICE* that
never ran. The version presented here draws substantially on work
by Marian Wilkinson in her book *The Carbon Club*. The rest of
the chapter, about the infrastructure about the global disinformation
campaign to stall action on climate change, rely on research by
Dr Jeremy Walker over multiple interviews and several archival
documents. Supplemental detail was provided by independent
sources.

The account of the exchange that opens Chapter 11 was produced
from a recording of the event, with supplemental information
provided by Ian Dunlop. The sections of the chapter exploring
why people continue to work for petroleum companies relies on
academic research identified through interviews with and assistance
from Dr Rob White, Dr Ellie Piggott, Dr Grace Augustine and
Dr Birthe Soppe. It also draws on my observations from interviews
I conducted with four former petroleum industry workers who are
not named in the chapter and from informal conversations with
other petroleum industry workers. The closing profile of 'Joshua'
demonstrates the ideas discussed in the previous section to show
how they actually play out, but the worker's name has been changed
to protect their identity as a condition of interview. His views are
his own.

The transcript of Tim Winton's speech that opens chapter 12 was
supplied to me as source material in previous news reporting. The

information about the tobacco industry and its activities is drawn from searches of the tobacco industry archive maintained at the University of California San Francisco and considerable research by the Centre for International Environmental Law. These documents are listed below. The description of William Walkley's interaction with Albert Namatjira is based on photos available via Trove and additional context provided by Dr Ruth Ellis. The outline of Walkley's sponsorship arrangements is based partly on research published by journalist Belinda Noble, supplemented with my own additional research. The final profile is a product of a two-hour interview with activist Joana Partyka and supplemented by documentary material listed below. This protest and the opposition in Western Australia are significant, given the first oil found in Australia was in Western Australia.

Chapter 13 opens with an account of a field visit to discuss the contrasting decisions involving the Gomeroi and Tiwi Islands First Nations people. It relies on direct observation, news accounts and other materials. Insight into the legal operation of the *Native Title Act* was provided by Dr Lily O'Neill, who wrote her PhD thesis on this subject. The sections on the Tiwi Islands were reconstructed from various accounts listed below, including field reporting by Lisa Cox of *Guardian Australia*. The biographical material about Dick Wells was taken from an industry history.

Chapter 14 is the product of hours of interviews, field work, direct observation and considerable previous reporting in *The Guardian Australia* and *The Saturday Paper*. These previously published stories are listed below. The section describing the speech by Misha Zelinsky is based on my own observations, as an accredited member of the media, at the session at which Zelinsky appeared at the Australian Domestic Gas Outlook conference in Sydney in 2022.

My account of the events described in chapter 15 was produced from material obtained while attending the 2023 APPEA conference as a registered member of the media, from my own news reporting for *The Guardian Australia* listed below, and from additional research undertaken around that time.

The final chapter contains my account after attending COP28

in Dubai as an accredited member of the media during the second week of the conferences. It summarises my direct observations and incorporates material that was published as reporting for *The Saturday Paper*. Aspects of the account relies on reporting from other journalists with *The Guardian*, *The BBC*, *The New York Times*, and *The Centre for Climate Reporting*. It draws from multiple interviews and conversations had during the seven days I was on the ground.

Source List

Prologue

Allam, L and Evershed, N 2019, 'Too hot for humans? First Nations people fear becoming Australia's first climate refugees' *The Guardian*, 18 December, <https://www.theguardian.com/australia-news/2019/dec/18/too-hot-for-humans-first-nations-people-fear-becoming-australias-first-climate-refugees#:~:text=Temperature%20records%20have%20already%20been,would%20not%20arrive%20until%202030>.

American Chemistry Society 2009, 'Development of the Pennsylvania oil industry', <https://www.acs.org/education/whatischemistry/landmarks/pennsylvaniaoilindustry.html>.

Arrhenius, S 1896, 'On the influence of carbonic acid in the air upon the temperature of the ground', *The London, Edinburgh and Dublin Philosophical Magazine and Journal of Science*, vol. 41, April, <https://www.rsc.org/images/Arrhenius1896_tcm18-173546.pdf>.

Arid Lands Environment Centre, 'The changing climate of Central Australia', 2023, <https://www.alec.org.au/the_changing_climate_of_central_australia>.

'Australian Displacement Updates', Internal Displacement Monitoring Centre, 2021, <https://www.internal-displacement.org/countries/australia>.

Cook, G, Dowdy A, Knauer J, Meyer M, Canadell P & Briggs P 2021, 'Australia's Black Summer of fire was not normal – and we can prove it', CSIRO, 29 November, <https://www.csiro.au/en/news/all/articles/2021/november/bushfires-linked-climate-change>.

'Climate & Perth', Water Corporation WA, 2024, <https://www.watercorporation.com.au/Our-water/Climate-change-and-WA/Climate-and-Perth#:~:text=Due%20to%20climate%20change%2C%20Perth's,around%2020%25%20since%20the%201970s.>.

'Climate Change 2021: The physical science basis', Working Group I contribution to the Sixth Assessment Report, IPCC, <chrome-extension://efaidnbmnnnibpcajpcglclefindmkaj/https://report.ipcc.ch/ar6/wg1/IPCC_AR6_WGI_FullReport.pdf>.

'Climate Change 2022': Mitigation of climate change', Working Group III contribution to the Sixth Assessment Report, IPCC, <https://www.ipcc.ch/report/ar6/wg3/

downloads/report/IPCC_AR6_WGIII_FullReport.pdf>.

'Climate Change 2023: Synthesis Report,' IPCC, <https://www.ipcc.ch/report/ar6/syr/downloads/report/IPCC_AR6_SYR_FullVolume.pdf >.

'Climate Change Roadmap Towards a Net-Zero and Resilient Future', Insurance Council of Australia, 2 November 2022, <https://insurancecouncil.com.au/resource/insurance-council-launches-insurer-roadmap-to-net-zero/>.

Davies, I 2022, Opening Address, 2022 APPEA Conference & Exhibition, 17 May, <https://energyproducers.au/all_news/speech-appea-chair-ian-davies-delivers-the-opening-address-of-the-2022-appea-conference-exhibition/>.

Dee, SG 2021, 'Scientists understood physics of climate change in the 1800s – thanks to a woman named Eunice Foote', *The Conversation*, 22 July, <https://theconversation.com/scientists-understood-physics-of-climate-change-in-the-1800s-thanks-to-a-woman-named-eunice-foote-164687>.

'Exxon memo on the Greenhouse Effect', 3 August 1988, <https://www.documentcloud.org/documents/3024180-1998-Exxon-Memo-on-the-Greenhouse-Effect>.

Fiebelkorn, A 2019, 'State Capture Analysis: How to quantitatively analyse the regulatory abuse by business-state relationships', (Discussion Paper no. 2), World Bank Group, June, <https://openknowledge.worldbank.org/server/api/core/bitstreams/c392604d-36d4-51b8-a82b-01b0b6a663aa/content>.

Frankfurt, H 1988, 'On Bullshit' in *The Importance of What We Care About*, Cambridge Univeristy Press, Cambridge, pp. 117–133, <https://www.cambridge.org/core/books/abs/importance-of-what-we-care-about/on-bullshit/F20BDF394713675F63385EA1823B9331>.

International Energy Agency 2021, 'Net-Zero by 2050: A roadmap for the global energy sector', May, <https://www.iea.org/reports/net-zero-by-2050>.

Jerving, S, Jennings, K, Hirsh, MM & Rust, S 2015, 'What Exxon knew about the Earth's melting Arctic', *Los Angeles Times*, 9 October, <https://graphics.latimes.com/exxon-arctic/>.

Kurmelovs, R 2021, 'The world's on fire, yet Australia keeps pumping out the gas', *VICE*, 11 August, <https://www.vice.com/en/article/88nbe4/australia-fossil-fuel-climate-emergency>.

Kurmelovs, R 2021, '"Like Snow": freak hail storms batter Australia's east coast', *The Guardian*, 20 October, <https://www.theguardian.com/australia-news/2021/oct/20/grapefruit-sized-australias-largest-hailstone-recorded-after-queensland-storms>.

Kurmelovs, R 2022, 'Australia spending billions on new gas pipelines that may end up worthless stranded assets', *The Guardian*, 23 February, <https://www.theguardian.com/australia-news/2022/feb/23/australia-spending-billions-on-new-gas-pipelines-that-may-end-up-worthless-stranded-assets>.

Kurmelovs, R 2022, 'Australia's oil and gas regulator criticised after chief hands out environmental "excellence" awards at industry dinner', *The Guardian*, 19 May, <https://www.theguardian.com/australia-news/2022/may/19/australias-oil-and-gas-regulator-criticised-after-chief-hands-out-environmental-excellence-awards-at-industry-dinner>.

Kurmelovs, R 2022, 'What the oil and gas industry tells itself,' *The Monthly*, July, <https://www.themonthly.com.au/issue/2022/july/royce-kurmelovs/what-oil-and-gas-industry-tells-itself#mtr>.

Kurmelovs, R 2022, 'The oil and gas industry refuses to die', *Rolling Stone Australia*, November, <https://au.rollingstone.com/culture/culture-features/the-oil-and-gas-industry-refuses-to-die-43310/>.

Kurmelovs, R & Tondorf, C 2022, 'Lismore flood: hundreds rescued and thousands evacuated as NSW city hit by worst flooding in history' *The Guardian*, 28 February, <https://www.theguardian.com/australia-news/2022/feb/28/lismore-flood-worst-history-nsw-floods-2022-rescued-evacuated>.

Ludlum, S 2021, *Full Circle*, Black Inc., Melbourne.

McGowan, M 2021, '"Heartbreaking"; Cyclone Seroja damages up to 70% of homes in West Australian town', *The Guardian*, 12 April, <https://www.theguardian.com/australia-news/2021/apr/12/cyclone-seroja-leaves-trail-of-damage-and-homes-without-power-in-western-australia>.

Milne, P 2022, '"This could be existential": Behind gas' desperate global game plan', *The Sydney Morning Herald*, 18 December 2022, <https://www.smh.com.au/business/companies/this-could-be-existential-behind-gas-desperate-global-game-plan-20221213-p5c60p.html>.

Semieniuk, G, Holden, BP, Mercure JF, et al. 2022, 'Stranded fossil-fuel assets translate to major losses for investors in advanced economies', *Nature Climate Change*, no. 12, 26 May, <https://www.nature.com/articles/s41558-022-01356-y>.

Shepherd, T 2023, 'Authorities struggle to deliver food and essentials to towns stranded by WA's "worst ever" floods' *The Guardian*, 6 January, <https://www.theguardian.com/australia-news/2023/jan/06/authorities-struggle-to-deliver-food-and-essentials-to-towns-stranded-by-was-worst-ever-floods>.

'Summary for Policymakers: Synthesis report of the IPCC Sixth Assessment Report, IPCC, <https://report.ipcc.ch/ar6syr/pdf/IPCC_AR6_SYR_SPM.pdf>.

Chapter 1

Australia, House of Representatives 2003, *Debates*, 'Environment: Kyoto Porocol', 2 December, <https://parlinfo.aph.gov.au/parlInfo/genpdf/chamber/hansardr/2003-12-02/0009/hansard_frag.pdf;fileType=application%2Fpdf>.

Baker, N 2019, 'Tony Abbott not convinced by "so-called" climate science', *SBSNews*, 12 April, <https://www.sbs.com.au/news/article/tony-abbott-not-convinced-by-so-called-climate-science/tnjskvhnt>.

Cook, G, Dowdy A, Knauer J, Meyer M, Canadell P & Briggs P 2021, 'Australia's Black Summer of fire was not normal – and we can prove it', CSIRO Publishing, Clayton, Victoria, 29 November, <https://www.csiro.au/en/news/all/articles/2021/november/bushfires-linked-climate-change>.

CSIRO 2022, 'Understanding the causes and impacts of flooding', CSIRO, <https://www.csiro.au/en/research/disasters/floods/causes-and-impacts>.

Fasullo, J, Rosenbloom, N & Buchholz R 2023, 'A multiyear tropical Pacific cooling response to recent Australian wildfires in CESM2', *Science Advances*, vol. 9, issue 19, 10 May, <https://www.science.org/doi/10.1126/sciadv.adg1213>.

Knaus, C & McGowan, M 2022, 'Climate change spat splits Lismore council in flood aftermath', *The Guardian*, 23 May, <https://www.theguardian.com/australia-news/2022/mar/23/climate-change-spat-splits-lismore-council-in-flood-aftermath>.

Kurmelovs, R 2023, 'Lismore one year after the floods,' *The Saturday Paper*, 4 March, <https://www.thesaturdaypaper.com.au/news/economy/2023/03/04/lismore-one-year-after-the-floods>.

Lerat, J, Vaze, J, Marvanek, S, Ticehurst, C & Wang, B 2022, 'Characterisation of the 2022 floods in the Northern Rivers region', CSIRO, 30 November, p. 6, <https://nema.gov.au/sites/default/files/inline-files/Characterisation%20of%20the%202022%20floods%20in%20the%20Northern%20Rivers%20region.pdf>.

Lu, D 2023, 'Black summer bushfires may have caused rare "triple dip" La Niña, study suggests', *The Guardian*, 11 May, <https://www.theguardian.com/australia-news/2023/may/11/black-summer-bushfires-may-have-caused-rare-triple-dip-la-nina-study-suggests>.

Morrison, G 2023, 'As Lismore slowly repairs it's preparing for higher floods under climate change' *Cosmos*, 24 February, <https://cosmosmagazine.com/earth/lismore-flood-climate-change-preparation/>.

Rennie, E 2022, 'Rahima's house burnt down in the Lismore floods – she's seen an opportunity,' *ABC News*, 11 March, <https://www.abc.net.au/news/2022-03-11/nsw-lismore-resident-sees-opportunity-amid-devastating-floods/100901758>.

Richmond River Historical Society, 'Lismore Chronology', <https://www.richhistory.org.au/lismore-history/lismore-chronology/>.

Ritchie, H & Roser, M 2020, 'CO$_2$ Emissions', Our World In Data, UK, June <https://ourworldindata.org/co2-emissions>.

Schmidt S 2023, 'El Niño and La Niña – what's it all about?', CSIRO, 20 September, <https://www.csiro.au/en/news/all/articles/2023/september/el-nino-la-nina#:~:text=Key%20points,to%20drought%20to%20extreme%20events>.

'Summary for Policymakers: Synthesis report of the IPCC Sixth Assessment Report, IPCC, <https://report.ipcc.ch/ar6syr/pdf/IPCC_AR6_SYR_SPM.pdf>.

Talberg, A, Hui, S & Loynes, K 2015, 'Australian climate change policy to November 2013: a chronology', Science, Technology, Environment and Resources Section, Parliament House Library, 9 Sepetmber, <https://www.aph.gov.au/About_Parliament/Parliamentary_Departments/Parliamentary_Library/pubs/rp/rp1516/ClimateChron>.

Chapter 2

Australian Associated Press 2018, 'Scott Morrison tells students striking over climate change to be "less activist"', *The Guardian*, 26 November, <https://www.theguardian.com/environment/2018/nov/26/scott-morrison-tells-students-striking-over-climate-change-to-be-less-activist>.

Australia, House of Representatives 2019, 'Petition EN1041 — Declare a Climate Emergency', 16 October, <https://www.aph.gov.au/e-petitions/petition/EN1041>.

Competition and Consumer Act 2010 (Cth) s 18.

Extinction Rebellion 2012, 'Canberra — XR Sets pram alight at Parliament', 10 August, (Facebook Video 13.23), <https://www.facebook.com/XRAustralia/videos/4242016295892653/?__tn__=%2CO-R>.

Francey, N & Chapman, S 2000 '"Operation Berkshire": the international tobacco companies' conspiracy', *British Medical Journal* 321:371, 5 August, <https://www.bmj.com/content/321/7257/371>.

Geoscience Australia 2022, 'Coal', Australian Government, <https://www.ga.gov.au/digital-publication/aecr2022/coal#:~:text=end%202020%20(Mt)-,Production,than%20200%20known%20coal%20deposits.>.

Gould, C 2021, 'Eight arrested after protesters spray paint Parliament House, set fire to pram' *News.com.au*, 10 August, <https://www.news.com.au/technology/environment/climate-change/climate-protesters-spray-paint-parliament-house/news-story/97575563c00f1b3f1788b2e8d2d9168e>.

Keoghan, S 2022, 'Flood-hit Lismore residents dump debris outside Kirribilli House', *The Sydney Morning Herald*, 21 March, <https://www.smh.com.au/national/nsw/flood-hit-lismore-residents-dump-debris-outside-kirribilli-house-20220321-p5a6dy.html>.

Kurmelovs, R 2021, 'Chalk paint and police raids: why climate activists are under fire' *The Guardian*, 4 December, <https://www.theguardian.com/environment/2021/dec/04/chalk-paint-and-police-raids-why-climate-activists-are-under-fire>.

Kurmelovs, R 2022, 'Fireproof Australia: who are the radical Extinction Rebellion splinter Group?' *The Guardian*, 10 April, < https://www.theguardian.com/environment/2022/apr/10/fireproof-australia-who-are-the-radical-extinction-rebellion-splinter-group>.

Milieudefensie et al. v Royal Dutch Shell plc (2021), C/09/571932 / HA ZA 19-379, <https://climatecasechart.com/wp-content/uploads/non-us-case-documents/2021/20210526_8918_judgment-1.pdf>.

Minister for the Environment v Sharma (2022) FCAFC 65 <https://www.judgments.fedcourt.gov.au/judgments/Judgments/fca/full/2022/2022fcafc0065>.

Municipalities of Puerto Rico v Exxon Mobil Corp. (2017), 3:22-cv-01550, <https://climatecasechart.com/case/municipalities-of-puerto-rico-v-exxon-mobil-corp/>.

Racketeer Influenced and Corrupt Organizations Act (RICO) 1970 (Congress), Public Law 91-452, <https://www.govtrack.us/congress/bills/91/s30/text>.

Re Australian Federation of Consumer Organisations Inc v the Tobacco Institute of Australia Limited (1991), FCA 17 – 12ATPR 41-079/98 ALR 670, <https://www8.austlii.edu.au/cgi-bin/viewdoc/au/cases/cth/federal_ct/1991/17.html>.

State of Minnesota v American Petroleum Institute (2020), 20-cv-01636, <https://climatecasechart.com/case/state-v-american-petroleum-institute/>.

Trade Practices Act 1974 (Cth), s 52.

Chapter 3

APEA, 1986, Petroleum Exploration and Development, Schools Resource Kit.

Australia, House of Representatives 1968, *Debates*, 'Price of Petrol and Petrol Products', 16 May, <https://parlinfo.aph.gov.au/parlInfo/search/display/display.w3p;query=Id%3A%22hansard80%2Fhansardr80%2F1968-05-16%2F0049%22>.

Avery, E 1960, APEA Meeting minutes, 24 February, in Wilkinson, R 2010, *Knights, Knaves And Dragons: 50 years inside APPEA and Australia's oil and gas politics,* Media Dynamics, Windsor, Queensland, pp. 52-53.

British Royal Commission 1912–14 *Fuel and Engines,* Report 'on the means of supply and storage of liquid fuel in peace and war, and its application to warship engines, whether indirectly or by internal combustion'.

Centre for International Environmental Law (CIEL), 2017, 'Smoke and Fumes: The legal and evidentiary basis for holding big oil accountable for the climate crisis', 16 November, <https://www.ciel.org/reports/smoke-and-fumes/>.

Columbia University Graduate School of Business & Nevins, A 1960 *Energy and Man: A Symposium* 1960, Appleton-Century-Crofts, University of Michigan, <https://www.documentcloud.org/documents/21094738-1959-energy-and-man-symposium #document/p16/a2061939>.

Franta, B 2022, 'Big Carbon's Strategic Response to Global Warming, 1950–2020', (Doctoral Dissertation), Standford University, p. 139, <https://stacks.stanford.edu/file/druid:hq437ph9153/Franta%20-%20Big%20Carbon%20strategic%20response%20 to%20global%20warming%201950-2020%20-%202022-08-25-augmented.pdf>.

Gardiner, B 2022, 'How an Early Oil Industry Study Became Key in Climate Lawsuits', *Yale Environment 360*, (Blog), Yale School of the Environment, 30 November, <https://e360.yale.edu/features/climate-lawsuits-oil-industry-research>.

Ikard, F 1965, 'Meeting the Challenges of 1966', *Proceedings of the American Petroleum Institute*, <https://www.climatefiles.com/trade-group/american-petroleum-institute/1965-api-president-meeting-the-challenges-of-1966/#:~:text=Ikard%20emphasized%20that%20 %E2%80%9C%5Bo%5D,local%20or%20even%20national%20efforts.%E2%80%9D>.

Johnstone, MH 1979, 'A Case History of Rough Range,' *The APPEA Journal, vol.* 19, CSIRO Publishing, Clayton, Victoria, pp. 1–6, <https://www.publish.csiro.au/AJ/AJ78001>.

Jones, CA 1958 'A review of the Air Pollution Research Program of the Smoke and Fumes Committee of the American Petroleum Institute', *Journal of the Air Pollution Control Association*, 8:3, pp. 268-272, <https://www.documentcloud.org/documents/2827789-1958-Charles-Jones-Smoke-and-Fumes-Committee-of.html#document/p4/a293900>.

Lipsky, D 2023, *The Parrot and the Igloo*, WW Norton & Company, New York, extracted in 'How Sun Myung Moon "Digested the Scientists" and Fueled Climate-Change Denial' *Rolling Stone*, 20 August, <https://www.rollingstone.com/politics/politics-features/sun-myung-moon-unification-church-climate-change-denial-1234808552/>.

O'Sullivan, T, McGarry, DJ, Kamenar, A, & Brown, RS 1991, 'The Discovery and Development of Moonie – Australia's first commercial oilfield', *APPEA Journal*, vol. 31(1), CSIRO Publishing, Clayton, Victoria, pp. 1–12, <https://www.publish.csiro.au/aj/aj90001>.

Ritchie, H & Roser, M 2020, 'CO_2 Emissions', Our World In Data, UK, June <https://ourworldindata.org/co2-emissions>.

Robinson, E & Robbins RC 1968, 'Sources, Abundance, and Fate of Atmospheric Pollutants', Report for the American Petroleum Institute, Stanford Research Institute, February, <https://www.smokeandfumes.org/documents/document16>.

Parry, J 2003, 'Jim Parry Reflects on the Rough Range Discovery' 2003, *Petroleum Exploration Society of Australia News,* no. 67, December, < https://archives.datapages.com/data/petroleum-exploration-society-of-australia/news/067/067001/pdfs/25.htm>.

Siller, B (Interview) in Wilkinson, R 2010, *Knights, Knaves And Dragons: 50 years inside*

APPEA and Australia's oil and gas politics, Media Dynamics, Windsor, Queensland, p. 35.

Sprigg, R 1959, 'Letter to the Editor: Australian Association of Exploration Independents', *Australasian Oil and Gas Journal,* vol. 5 no. 7, p. 10.

Stanford Research Institute, *Institute Sponsored Research: January 1951 through June 1952,* Booklet, Stanford, California, p. 1 <https://insideclimatenews.org/wp-content/uploads/2016/06/SRI-Booklet-1951-1952.pdf>.

White House 1965, 'Restoring the Quality of Our Environment', Report of the Environmental Pollution Panel President's Science Advisory Committee, November, <https://www.climatefiles.com/climate-change-evidence/presidents-report-atmospher-carbon-dioxide/>.

Wilkinson, R 2010, *Knights, Knaves And Dragons: 50 years inside APPEA and Australia's oil and gas politics,* Media Dynamics, Windsor, Queensland pp. 19–23, 32–34, 37–39, 41–44, 46–57, 140, 407–411.

Chapter 4

'Air pollution blamed on vehicles', *The Canberra Times,* 4 April 1972, <https://trove.nla.gov.au/newspaper/article/102207286?searchTerm=senate%20select%20committee%20air%20pollution>.

APEA, 1986, Petroleum Exploration and Development, Schools Resource Kit.

Australian Academy of Science Committee on Climate Change, 1976, *Report of a Committte on Climatic Change.*

Australia, Royal Commission into Exploratory and Production Drilling for Petroleum in the Area of the Great Barrier Reef, Final Report, 1974, Australian Government Publishing Service, Canberra.

Chevron Global, 'The Gorgon Project', <https://australia.chevron.com/our-businesses/gorgon-project>.

Clarke, KC & Hemphill, JJ, 'The Santa Barbara Oil Spill: A retrospective' in Danta, D (ed) 2002 *Yearbook of the Association of Pacific Coast Geographers,* University of Hawaii Press, vol. 64, pp. 157–59, <https://people.geog.ucsb.edu/~kclarke/Papers/SBOilSpill1969.pdf>.

Clean Air Society of Australia and New Zealand 1972, *Clean Air Conference: Proceedings of the Clean Air Conference,* Melbourne.

Australia, Senate 1977, *Debates,* Senate, Senator Mulvihill, J Senate Standing Committee on Science and the Environment, 25 August, <https://parlinfo.aph.gov.au/parlInfo/search/display/display.w3p;db=HANSARD80;id=hansard80%2Fhansards80%2F1977-08-25%2F0093;query=Id%3A%22hansard80%2Fhansards80%2F1977-08-25%2F0013%22>.

Connolly, W 1973, 'World Energy Growth' *APEA Journal,* vol 13(2), CSIRO Publishing, Clayton, Victoria, <https://chrome-extension://efaidnbmnnnibpcajpcglclefindmkaj/https://www.publish.csiro.au/ep/pdf/AJ72029>.

'Death of Professor Harry Bloom', *Australian Jewish News,* 16 October 1992, < https://trove.nla.gov.au/newspaper/article/261606396?searchTerm=Professor%20Harry%20Bloom>.

Dundon, R 2018, 'Photos: LA's mid-century smog was so bad, people thought it was a gas attack', *Medium,* 23 May, <https://medium.com/timeline/la-smog-pollution-4ca4bc0cc95d>.

'Expensive spectacle,' *The Bulletin*, 14 December 1968, <https://nla.gov.au/nla.obj-674387366/view?sectionId=nla.obj-691244359&searchTerm=Langston+Esso&partId=nla.obj-674556085#page/n17/mode/1up/search/Langston+Esso>.

Hartmann, H et al 1972, *Nature In The Balance*, Heinemann Educational Books, London.

Hartmann, H 1973, 'The Role of Gas in Environmental Control' *APEA Journal*, vol. 13(10), CSIRO Publishing, Clayton, Victoria, pp. 125–131, <https://www.publish.csiro.au/AJ/AJ72019>.

Langston, JD 1970, 'Petroleum and the Environment' *APEA Journal*, vol. 10(1), CSIRO Publishing, Clayton, Victoria, pp. 16–18.

Marchant, S 2010, Parliament of Australia, *The Biographical Dictionary of the Australian Senate*, Sir Condor Louis Laucke (1914–1993),<https://biography.senate.gov.au/laucke-condor-louis/>.

Murray, R & Thomas, T 2014, 'Bob Foster: Petroleum scientist and individualist who never feared a controversy', *The Sydney Morning Herald*, 6 July, <https://www.smh.com.au/national/bob-foster-petroleum-scientist-and-individualist-who-never-feared-a-controversy-20140704-zswef.html>.

Nelson, K 1971, 'Clean Air' (Speech), 1971 National Seminar, Australian Mining Industry Council, 19 April, p. 3.

'"No apology" for Esso-BHP profit', *The Canberra Times*, 4 December 1968, <https://trove.nla.gov.au/newspaper/article/136959145?searchTerm=Langston%20Esso>.

'Obituary 3: Dr William H Lang', *The New York Times*, 31 December 1970, <https://www.nytimes.com/1970/12/31/archives/obituary-3-no-title.html>.

'Oil and Gas Industry Environmental Control Expert in Melbourne', 1970, *The Australian Gas Journal*, no. 24.

Pendal, P 2010, Parliament of Australia, *Biographical Dictionary of the Australian Senate*, George Howard Branson (1918–1999), <https://biography.senate.gov.au/branson-george-howard/>.

Ritchie, H & Roser, M 2020, 'CO_2 Emissions', Our World In Data, UK, June <https://ourworldindata.org/co2-emissions>.

Senate Select Committee 1969, *Air Pollution: Report Part I*, Parliamentary Paper no. 91, 10 September, p. 4, <https://parlinfo.aph.gov.au/parlInfo/search/display/display.w3p;query=Id%3A%22publications%2Ftabledpapers%2FHPP032016005867_4%22>.

Senate Select Committee 1969, *Air Pollution: Report Part II*, Minutes of Evidence, Parliamentary Paper no. 91, 10 September, pp. 184, 644–649.

Sherwin, RJS 1972, 'Energy – Major Sources and Consumption', *APEA Journal,* vol. 12(1), CSIRO Publishing, Clayton, Victoria, p. 103, <https://www.publish.csiro.au/aj/aj71018>.

Sullivan, M 2007, 'Contested Science and Exposed Workers: ASARCO and the occupational standard for inorganic arsenic', *Public Health Report*, vol. 122, issue 4, July–August, <https://www.ncbi.nlm.nih.gov/pmc/articles/PMC1888505/>.

Trumbull, R 1970, 'Australia Postpones Oil-Drilling Plan' *The New York Times,* 25 June, <https://www.nytimes.com/1970/01/25/archives/australia-postpones-oildrilling-plan.html>.

Vaughan, A 2017, 'Torrey Canyon disaster – the UK's worst-ever oil spill 50 years on', *The Guardian*, 18 March, <https://www.theguardian.com/environment/2017/mar/18/

torrey-canyon-disaster-uk-worst-ever-oil-spill-50tha-anniversary>.

Walker S 1974, Interview with Sir Henty Denham, (Sound Recording), 2 December, <https://nla.gov.au/nla.obj-215011730/listen>.

Wilkinson, R 2010, *Knights, Knaves And Dragons: 50 Years Inside APPEA and Australia's Oil and Gas Politics*, Media Dynamics, Windsor, Queensland, pp 67–68, 82–86.

Chapter 5

Gwynne, P 2014, 'My 1975 "cooling world" story doesn't make today's climate scientists wrong' *Inside Science*, 21 May, <https://www.insidescience.org/news/my-1975-cooling-world-story-doesnt-make-todays-climate-scientists-wrong>.

Maloney, S & Grosz, C 2023, 'Rex Connor and Tirath Khemlani' *The Monthly*, October, <https://www.themonthly.com.au/issue/2013/october/1380549600/shane-maloney-and-chris-grosz/rex-connor-and-tirath-khemlani#mtr>.

Peterson, TC, Connolley, WM & Fleck, J 2008, 'The myth of the 1970s global cooling scientific consensus', *Bulletin of the American Meteorological Society*, vol. 89, issue 9, September, <https://nora.nerc.ac.uk/id/eprint/11584/1/2008bams2370%252E1.pdf>.

World Meteorological Organization 1975, 'Proceedings of WMO/IAMAP Symposium on long-term climatic fluctuations, 18–23 August', <https://library.wmo.int/index.php?lvl=notice_display&id=7449>.

Sprigg, R 1977, 'Towards an APEA Code of Environmental Practice for the Onshore Petroleum Industry', *APEA Journal*, vol. 37, CSIRO Publishing, Clayton, Victoria.

Toohey, B 2000, 'The affair that led to Whitlam's sacking', *The Australian Financial Review*, 11 November, <https://www.afr.com/politics/the-affair-that-led-to-whitlams-sacking-20001111-k9u99>.

Weidenbach, K 2008, *Rock Star: the story of Reg Sprigg – an outback legend*, Midnight Sun Publishing, Adelaide.

Weidenbach, K 2014, *Blue Flames, Black Gold: the story of Santos*, Santos Pty Ltd, Adelaide.

Wilkinson, R 2010, *Knights, Knaves And Dragons: 50 Years Inside APPEA and Australia's Oil and Gas Politics*, Media Dynamics, Windsor, Queensland, pp 119, 122, 166–171.

Chapter 6

Abel B, 1980, Letter to Barker, L, confirming commissions and advising, APEA, 30 May.

APEA 1980, 'About Noonkanbah And The Sacred Sites Movement in Western Australia: A Composite Report', APEA.

APEA 1980, Minutes of Council Meeting with Ayers, AJ, Secretary Department of Aboriginal Affairs, 27 August.

APEA 1980, 'Noonkanbah: A report on events', APEA.

APEA 1980, 'Notes on Discussion Re: Aboriginal Sacred Sites Issue', 27 August.

APEA 1980, 'Summary of meeting on Aboriginal Affairs', non-verbatim summary with AMIC and senior ministers, 28 August.

Barker, L 1980, 'Aboriginal Land Rights: The Commonwealth, Northern Territory, Queensland, New South Wales, Victoria and South Australia', APEA.

Barker L 1980, Letter to Abel, B, regarding commission, parameters of brief and experience, 19 May.

Barker, L 1980, Parameters of study (annotated), 19 May.

Barker L 1980, Letter to Ted Garland regarding Aboriginal Land Rights Study and payment, 17 June.

Barker L, 1980, Letter to Frank Lane-Mullins regarding file cards and background material, 4 July.

Barker L, 1980, Letter to Ted Garland regarding progress, 25 July.

Barker L, 1980, Letter to Joseph Poprzeczny regarding instructions on trip to Kimberleys, 2 August.

Barker L, 1980, Letter to Bob Liddle regarding meeting, 20 August.

Barker L, 1980, Letter to Ted Garland regarding fee and size of team, 26 August.

Barker, L 1980, Handwritten notes of meeting in parliament, 28 August.

Barker L 1980, 'Possible Joint Statement by Deputy Prime Minister, AMIC and APEA draft media release, 28 August.

Barker L 1980, Letter to Joseph Poprzeczny regarding commission, 17 September.

Barker, L 1980, Notes form meeting with Fred Chaney, 18 September.

Barker L 1980, Letter to Eric Webb regarding completion of two reports and embedded researcher, 18 December.

Barker L 1980, Letter to Ted Garland regarding reports on the "aboriginal problem", 18 December.

Barker L 1981, Letter to Eric Webb regarding additional speech, 6 January.

Barker L 1981, Letter to Keith Orchison regarding payment, 30 January.

Benbow D 1980, Letter to Len Barker regarding permission to publish, 15 July.

Biskup P 1973, *Not Slaves, Not Citizens: the Aboriginal problem in Western Australia, 1898–1954*, UQP, St Lucia, Queensland.

Hawke, S & Gallagher, M 1989, *Noonkanbah: Whose land, whose law*, Fremantle Press, Freemantle, Western Australia.

Jamieson, R 2011, *Charles Court: I love this place*, Fremantle Press, Freemantle, Western Australia.

Jones P, Letter to Orchison, K, regarding guidelines, 13 February.

Kolig, E 1987, *The Noonkanbah Story*, University of Otago Press, Dunedin, New Zealand.

Kurmelovs, R 2017, 'The forgotten Australian war behind this harrowing photo' *VICE*, 20 February, <https://www.vice.com/en/article/vvxm9m/the-forgotten-australian-war-behind-this-harrowing-photo>.

Lane-Mullins F 1980, Letter to Barker, L, regarding article by Bishop Jobst, 18 September.

Marra Worra Worra Aboriginal Corporation, 'Our Communities: Yungngora', <https://mww.org.au/our-communities/yungngora/>.

O'Neill, L 2016, 'A Tale of Two Agreements: Negotiating Aboriginal land access agreements in Australia's natural gas industry', (Doctoral Thesis), University of Melbourne, Melbourne.

Orchison K 1981, Letter to Barker, L regarding Aboriginal Land Rights Study, payment and late delivery, 29 January

Orchison, K 1981, letter to APEA Committee re: Advice to members on contact with Aboriginal communities, 13 July.

Orchison, K 1981, 'Aboriginal communities and petroleum exploration and development: Advice for members of the Australian Petroleum Exploration Association Limited', July.

Patterson K 1980, Letter to Barker, L, regarding statements by Northern Territory Lands Council, 18 September.

Poprzeczny J 1980, Letter to Barker, L, regarding delays, draft reports future meetings, 9 July.

Poprzeczny J 1980, Letter to Barker, L regarding completion of trip to Kimberley and Hawke Comments, August.

Poprzeczny J 1980, Letter to Barker, L regarding unused travel tickets from trip and thoughts, 25 August.

Poprzeczny J 1980, Letter to Len Barker regarding student unions, update and activists, 23 September.

Poprzeczny J (undated), Letter to Barker, L regarding completion and oil industry "programme".

Stevenson T, Letter to Len Barker regarding assignment and confirming brief, 17 June.

Reynolds JO 1980, Letter to O Bavinton regarding introduction of Len Barker, 26 June.

Richards, K 1980, Letter to Len Barker regarding extract from meeting minutes, 30 July.

Richardson, J 2020, '"They couldn't break me": Don McLeod, champion for Aboriginal justice in the Pilbara', (Doctoral Thesis), Monash University, Melbourne.

Vassiley, A 2021, 'Noonkanbah 1979: when unionists stood up for Aboriginal land rights', Red Flag, 17 February, <https://redflag.org.au/node/7540>.

Wilkinson, R 2010, Knights, Knaves And Dragons: 50 Years Inside APPEA and Australia's Oil and Gas Politics, Media Dynamics, Windsor, Queensland, pp 185–187, 189–192.

Chapter 7

APEA 1984, 'External Liaison and Information Services', APEA Annual Report.

APEA 1989, 'Atmospheric Emissions', APEA Annual Report.

Australian Institute of Petroleum 1981, Why keep lead in petrol?, (Pamphlet) The Institurte, Melbourne.

Australian Institute of Petroleum 1983, 'Director's Report', AIP Annual Report.

Australian Institute of Petroleum 1984, 'External Liaison and Information Services'", AIP Annual Report.

Australian Institute of Petroleum 1989, 'Director's Report', AIP Annual Report, 1989.

Australian Institute of Petroleum 1989, 'The Greenhouse Effect: A position paper', AIP, July.

Black, J 1978, 'The Greenhouse Effect', Presentation to Exxon Corporation Management Committee, (Transcript), 18 May, <https://insideclimatenews.org/wp-content/uploads/2015/09/James-Black-1977-Presentation.pdf>.

Bonneuil, C, Choquet, PL & Franta, B 1971, 'Early warnings and emerging accountability: Total's responses to global warming, 1971–2021', Global Environmental Change, vol. 71, November, <https://www.sciencedirect.com/science/article/pii/S0959378021001655>.

Bryce, M 1983, 'Incentives for Exploration', APEA Journal, vol. 23(2), CSIRO Publishing, Clayton, Victoria, p. 11.

Chui, G 1988, 'Circumstantial case for the greenhouse effect' The Canberra Times, 26 March, <https://trove.nla.gov.au/newspaper/article/101984809?searchTerm=James%20Hansen%20NASA>.

'"Dire consequences' in global warm-up', *The Australian*, 20 October 1983.

Dembicki, G 2023, *The Petroleum Papers: Inside the far-right conspiracy to cover up climate change,* Greystone Books, Vancouver.

Franta, B 2022, 'Big Carbon's Strategic Response to Global Warming, 1950-2020', (Dissertation), Standford University, August, p.141 <https://stacks.stanford.edu/file/druid:hq437ph9153/Franta%20-%20Big%20Carbon%20strategic%20response%20to%20global%20warming%201950-2020%20-%202022-08-25-augmented.pdf>.

'Fossil Fuels and the Greenhouse Effect' 1981, (Report), Office of National Assessments.

Global Climate Coalition Membership 1989, *Climate Files,* <https://www.climatefiles.com/denial-groups/global-climate-coalition-collection/1989-membership/>.

Hansen, J 1988, 'Hearing Before the Committee on Energy and Natural Resources, United States Senate, One Hundredth Congress, First Session on the Greenhouse Effect and Climate Change, Part 2, 23 June, <https://babel.hathitrust.org/cgi/pt?id=uc1.b5127807&view=1up&seq=45>.

Heath, DP 1980, *Choices and Challenges*: Papers delivered in the enviornmental protection technical stream session of the AIP's 1980 Congress on Future of Petroleum in the Pacific Region, Sydney, September 14, AIP, Melbourne.

Hudson, M 2015, '25 years ago the Australian government promised deep emission cuts, and yet here we still are' *The Conversation,* 9 October, <https://theconversation.com/25-years-ago-the-australian-government-promised-deep-emissions-cuts-and-yet-here-we-still-are-46805>.

Kelly, Ros 1990, 'Government sets target for reductions in greenhouse gases', Australian Government joint statement media release, 11 October, <https://parlinfo.aph.gov.au/parlInfo/search/display/display.w3p;query=Id:%22media/pressrel/2209922%22>.

Kroll, S 2012, *Private Empire: ExxonMobil and American Power,* Penguin Books, New York.

Lemlin, JS & Graham-Bryce, 1994, 'The petroleum industry's response to climate change: The role of the IPIECA Global Climate Change Working Group', *Industry and Environment,* vol. 17(1), pp. 27–30.

Matthews, C 2023, 'Inside Exxon's Strategy to Downplay Climate Change' *The Wall Street Journal,* 14 September 2023, <https://www.wsj.com/business/energy-oil/exxon-climate-change-documents-e2e9e6af>.

Readfern, G 2021, 'Australia's spy agency predicted the climate crisis 40 years ago – and fretted about coal exports' *The Guardian,* 28 November, <https://www.theguardian.com/environment/2021/nov/28/australias-spy-agency-predicted-the-climate-crisis-40-years-ago-and-fretted-about-coal-exports>.

Richardson, J 2020, '"They couldn't break me": Don McLeod, champion for Aboriginal justice in the Pilbara', (Doctoral Thesis), Monash University, Melbourne.

Ritchie, H & Roser, M 2020, 'CO$_2$ Emissions', Our World In Data, UK, June <https://ourworldindata.org/co2-emissions>.

Stephens, T 2009, 'Mining leader with motor oil in his veins' *The Sydney Morning Herald,* 16 January, <https://www.smh.com.au/national/mining-leader-with-motor-oil-in-his-veins-20090115-7hxc.html>.

Supran, G, Rahmstorf, S & Oreskes, N, 'Assessing ExxonMobil's global warming projections', *Science,* vol. 379, issue 6628, 13 January, <https://www.science.org/doi/10.1126/science.abk0063>.

The Financial Times 1989, 'The clouded view of the global greenhouse' in *The Canberra Times*, 30 June, <https://trove.nla.gov.au/newspaper/article/122273405?searchTerm=James%20Hansen%20NASA>.

Wilkinson, R 2010, *Knights, Knaves And Dragons: 50 Years Inside APPEA and Australia's Oil and Gas Politics*, Media Dynamics, Winton, Queensland, pp. 160–162, 175–179,194, 210, 247–250, 269–275, 281.

Chapter 8

APEA, 1986, *Petroleum Exploration and Development*, Schools Resource Kit.

Battrick, MA, Bishop, BR & Edmondson, GA 2004, 'Engaging the education community through petroleum industry sponsored programs', *APPEA Journal*, vol. 41(1), CSIRO Publishing, Clayton, Victoria, pp. 821–834.

Boon, HJ 2016, 'Pre-Service Teachers and Climate Change: A Stalemate', *Australian Journal of Teacher Education*, vol. 41, issue 4, <https://ro.ecu.edu.au/cgi/viewcontent.cgi?article=3075&context=ajte>.

Ghamgosar, N, Holmstrom, M, Jones, D & Terry, D 2022 'Our Workforce for the Future', (Event), APPEA Conference, Brisbane, 19 May.

Grossman, E 1994, 'Industry and Education – A Partnership for Learning' *APEA Journal*, vol. 34(1), CSIRO Publishing, Clayton, Victoria, pp. 853–861.

Ferguson, A 2018, 'Dollarmites Bites: the scandal behind the Commonwealth Bank's junior savings program', *The Sydney Morning Herald*, 18 May, <https://www.smh.com.au/business/banking-and-finance/dollarmites-bites-the-scandal-behind-the-commonwealth-bank-s-junior-savings-program-20180517-p4zfyr.html>.

Hudson, M 2023, 'Australian films *The Coal Question* and *What To Do About CO_2?*' All Our Yesterdays, 1 February, <https://allouryesterdays.info/2023/02/01/australian-films-the-coal-question-and-what-to-do-about-c02-interview-with-russell-porter/>.

Kurmelovs, R 2022, 'Fossil Fuels in Schools: Industry faces pushback in fight for hearts and minds of next generation', *The Guardian*, 4 October, <https://www.theguardian.com/australia-news/2022/oct/04/fossil-fuels-in-schools-industry-faces-pushback-in-fight-for-hearts-and-minds-of-next-generation>.

Truth Tobacco Industry Documents, economics professors, University of California, San Francisco, <https://www.industrydocuments.ucsf.edu/tobacco/docs/#id=ptvh0047>.

Truth Tabacco Industry Documents 1987, Memo from Dyer DM to Minshew G, 'Economic Witness Evaluation', University of California, San Francisco, <https://www.industrydocuments.ucsf.edu/tobacco/docs/#id=flhh0034>.

McGregor, P 1993, *The Australian Petroleum Resource Book*, AIP.

'Oil: Facts about the industry', (Poster), 1994, APEA.

Queensland State Government, 'Gateway to Industry Schools Program', <https://desbt.qld.gov.au/training/employers/gateway-schools>.

Queensland Minerals and Energy Academy, 'About us', <https://qmea.org.au/about-us-2/>.

Saving, TR 1999, 'S Charles Maurice: In Memoriam', *Southern Economics Association*, vol. 66, issue 2, October, <https://onlinelibrary.wiley.com/doi/epdf/10.1002/j.2325-8012.1999.tb00270.x>.

The Big Book of Oil And Gas (c. 1997), AIP–APPEA and AGA, Educational Communications.

The Coal Question, 1982 (Documentray), directed by Russell Porter, CEO.

The story of oil 1984, Pamphlet, AIP.

@Timinclimate 2022, Questacon video about energy transition, 7 July, <https://twitter.com/timinclimate/status/1412710532570697731?s=20>.

Truth Tobacco Industry Doucments 1984, 'The Tobacco Institute Cigarette Excise Tax Plan', 30 April, p. 109, <https://www.industrydocuments.ucsf.edu/tobacco/docs/#id=nfyh0051>.

Tytler, R & Freebody, P 2023, 'How should we teach climate change in schools? It starts with "turbo charging" teacher education" *The Conversation*, 13 June, <https://theconversation.com/how-should-we-teach-climate-change-in-schools-it-starts-with-turbo-charging-teacher-education-207221>.

Verral, M, Judd, K, O'Neill, M, Pugh, D & Snelling L 2022, 'Energy Leader's Panel', (Event), APPEA Conference, Brisbane, 17 May.

What to do about CO$_2$? 1984, (Documentary), directed by Russell Porter, CSIRO, <https://csiropedia.csiro.au/what-to-do-about-co2-1984/>.

Wilkinson, P 2019, 'David Bellamy obituary' *The Guardian*, 13 December, <https://www.theguardian.com/environment/2019/dec/12/david-bellamy-obituary>.

Wilkinson, R 2010, *Knights, Knaves And Dragons: 50 Years Inside APPEA and Australia's Oil and Gas Politics*, Media Dynamics Australia, Winton, Queensland, pp. 74, 221, 248, 285, 334–335.

Chapter 9

Australian Government Department of Industry, Science, Energy and Resources 2021, *Australia's Long-Term Emissions Reduction Plan: A whole-of-economy Plan to achieve net zero emissions by 2050*, <https://www.dcceew.gov.au/sites/default/files/documents/australias-long-term-emissions-reduction-plan.pdf>.

Australian Industry Greenhouse Network Annual Report 2021–22, AIGN, <https://www.aign.net.au/documents/AIG2495_Annual%20Report_Online_FA.pdf>.

Baxter, C 2012, 'Exxon's Tillerson: Could we really have expected a tiger to change its stripes? *DeSmog*, 28 June, <https://www.desmog.com/2012/06/28/exxon-s-tillerson-could-we-really-have-expected-tiger-change-its-stripes/>.

Beckett, C 2006, 'Reg Sprigg Medal – Barry Jones', *APPEA Journal*, CSIRO Publishing, Clayton, Victoria.

Bonneuil, C, Choquet, PL & Franta, B 2021, 'Early Warnings and Emerging Accountability: Total's responses to global warming, 1971–2021, *Global Environmental Change*, vol. 71, November, <https://www.sciencedirect.com/science/article/pii/S0959378021001655>.

Commonwealth of Australia 1997, 'In The National Interest: Australia's Foreign and Trade Policy', (White Paper), <http://repository.jeffmalone.org/files/foreign/In_the_National_Interest.pdf>.

Report of the Senate Environment, Communications, Information Technology and the Arts References Committee 2000, *The Heat Is On: Australia's Greenhouse Future*, 7 November, <https://www.aph.gov.au/parliamentary_business/committees/senate/

environment_and_communications/completed_inquiries/1999-02/gobalwarm/report/c08a>.

Gocher, D 2021, 'Gaslighting: How APPEA and its members continue to oppose genuine climate action', Australian Centre for Corporate Responsibility, June, <https://www.accr.org.au/downloads/2021-06-accr-appea-gaslighting-report.pdf>.

Goldenberg, S 2015, 'Exxon Knew of climate change in 1981, email says – but it funded deniers for 27 more years' *The Guardian*, 9 July, <https://www.theguardian.com/environment/2015/jul/08/exxon-climate-change-1981-climate-denier-funding>.

Leggett, J 2001, *Carbon War: Global warming and the end of the oil era*, Routledge, New York.

Hamilton, C 2006, 'The Dirty Politics of Climate Change', (Speech), Climate Change and Business Conference, Hilton Hotel, Adelaide, 20 February.

Hamilton, C 2007, *Scorcher: The Dirty Politics of Climate Change*, Black Inc, Melbourne.

Hodder, R 2011, 'Climate conflict: Players and tactics in the greenhouse game', (Doctoral Thesis), University of Wollongong, Wollongong, <https://ro.uow.edu.au/cgi/viewcontent.cgi?article=4618&context=theses>.

Hudson, M 2017, 'Cabinet papers 199–93: Australia reluctant while world moves towards first climate treaty' *The Conversation*, 1 January, <https://theconversation.com/cabinet-papers-1992-93-australia-reluctant-while-world-moves-towards-first-climate-treaty-70535>.

Hudson, M 2018, 'Enacted inertia: incumbent resistance to carbon pricing in Australia, 1989–2011', (Doctoral Thesis), University of Manchester, Manchester, <https://research.manchester.ac.uk/en/studentTheses/enacted-inertia-incumbent-resistance-to-carbon-pricing-in-austral>.

Jones, B 1997, 'Greenhouse policy – a quick brief', APPEA Report, member newsletter, no. 754, 12 July, p. 5.

King, E 2016, '25 Years Later: When the world united against climate change', *Climate Home News*, 3 July, <https://www.climatechangenews.com/2016/03/07/25-years-later-when-the-world-united-against-climate-change/>.

Mabo v. Queensland (No 2) (1992) HCA 23, 175 CLR 1, < http://www8.austlii.edu.au/cgi-bin/viewdoc/au/cases/cth/HCA/1992/23.html?stem=0%3D0%3DMabo%20175%20CLR>

Martin, S 2021, 'Scott Morrison refuses to release net zero 2050 modelling amid condemnation of climate policy' *The Guardian*, 27 October, <https://www.dcceew.gov.au/sites/default/files/documents/australias-long-term-emissions-reduction-plan.pdf>.

McGregor, I 2008, *Organising to influence the Global Politics of Climate Change*, (Research), University of Technology Sydney, Sydney, <https://opus.lib.uts.edu.au/bitstream/10453/11492/1/2008000811OK.pdf>.

Morrison, S & Taylor, A 2021, 'Australia's Plan to Reach Our Net Zero Target by 2050', Australian Government joint statement media release, 26 October, <https://www.minister.industry.gov.au/ministers/taylor/media-releases/australias-plan-reach-our-net-zero-target-2050>.

Montague, B 2015, 'How Clinton's Presidency Caused Oil-Funded Lobbyists to Intensify Attacks on Climate Science', *DeSmog*, 28 January, <https://www.desmog.com/2015/01/28/how-clinton-presidency-caused-oil-funded-lobbyists-intensify-attacks-climate-science/>.

Pearse, G 2005, 'The Business Response to Climate Change: Case studies of Australian interest groups', (Doctoral Thesis), Australian National University, <https://openresearch-repository.anu.edu.au/bitstream/1885/109792/4/b22560233-pearse_g.pdf>.

Pearse, G 2007, *High & Dry*, Viking, Sydney.

Ritchie, H & Roser, M 2020, 'CO_2 Emissions', Our World In Data, UK, June <https://ourworldindata.org/co2-emissions>.

'The Greenhouse Challenge / Challenge Plus – Industry Partnerships' *International Energy Agency*, (Program), 5 November 2017, <https://www.iea.org/policies/481-the-greenhouse-challenge-challenge-plus-industry-partnerships>.

'WA Businessman Helped Bankroll Abbott Fund', *ABC Online*, 28 August 2003, <https://www.abc.net.au/news/2003-08-28/wa-businessman-helped-bankroll-abbott-fund/1471048>.

Wilkinson, M 2020, *The Carbon Club*, Allen & Unwin, Sydney.

Wilkinson, R 2010, *Knights, Knaves And Dragons: 50 Yeas Inside APPEA and Australia's Oil and Gas Politics*, Media Dynamics, Winton, Queensland, pp. 304–308, 327.

Worden, S 1998, 'The Case Against the Carbon Tax: The industry greenhouse network's 1994/95 campaign', (Masters Thesis), Queensland University of Technology, Brisbane.

Chapter 10

Australian Institute of Petroleum 1986, 'Position Paper for Energy 2000', 3–4 September.

Australian Institute of Petroleum 1986, 'Petroleum Taxation: The need for reform: a response to questions raised by the Minister for Resources and Energy at the Energy 2000 Conference', November.

Bonython, J 1973, 'Statement' in Connor Records, Series 1, D61, Box 209, Santos Pty Ltd, Adelaide, 8 November.

Bonython, J 1976, Letter to Murchison, J, regarding introduction of Fisher, A, in the Hoover Institute archive by Walker, J, 19 August.

Brown, JD 1963, 'A Big Man Even In Big D', *Sports Illustrated*, 21 January, <https://vault.si.com/vault/1963/01/21/a-big-man-even-in-big-d>.

Cartwright, G 1987, 'Texas Primer: Clint Murchison, Jr.', *Texas Monthly*, June, <https://www.texasmonthly.com/arts-entertainment/texas-primer-clint-murchison-jr/>.

Christie, A 1986, 'BP Group in Australia Submission to Energy 2000 National Energy Policy Review', BP Group, July.

Commonwealth Department of Resources and Energy 1986, *Energy 2000: A National Energy Policy Review*, March.

Commonwealth 1989, Submission No 6363 – Australian Response to the Greenhouse Effect and Related Climate Change, (Cabinet Minute), no. 12416, 3 April, <https://recordsearch.naa.gov.au/SearchNRetrieve/Interface/ViewImage.aspx?B=31430608>.

Commonwealth Department of Primary Industries and Energy 1988, 'Energy 2000: A National Energy Policy Paper, (Policy Paper), no. 102, Australian Government Publishing Services, Canberra, <https://nla.gov.au/nla.obj-2518051825/view?searchTerm=Energy+2000&partId=nla.obj-2520690372#>.

Commonwealth Department of Prime Minister and Cabinet 1989, *Greenhouse Effect – International Action*, (Watching brief), no. 3820, 30 August, <https://

recordsearch.naa.gov.au/SearchNRetrieve/Interface/ViewImage.aspx?B=7903688>.

'Competitive Enterprise Institute' Climate Investigations Center, <https://climateinvestigations.org/competitive-enterprise-institute/>.

Curtis, A 2011, 'The Curse of Tina' *BBC* , 13 September, <https://www.bbc.co.uk/blogs/adamcurtis/entries/fdb484c8-99a1-32a3-83be-20108374b985>.

Dickinson, T 2014, 'Inside the Koch brothers' toxic empire', *Rolling Stone*, 14 September, <https://www.rollingstone.com/politics/politics-newsinside-the-koch-brothers-toxic-empire-164403/>.

Exxon 1988, 'The Greenhouse Effect', (Memo), <https://www.documentcloud.org/documents/3024180-1998-Exxon-Memo-on-the-Greenhouse-Effect>.

Esso 1986, '"Energy 2000" A National Energy Policy Review Comments by Esso Australia, (Policy Submission), July.

Foster, B 2006, 'Editorial', *Energy & Environment*, vol. 17, no. 1. January, <https://www.jstor.org/stable/44397024>.

Gifford, RM 1986, 'Submission to Energy 2000', (Policy Submission), May.

Guilliatt, R 2016, 'True Disbeliever: Senator Malcolm Roberts' crusade', *The Australian*, 22 October, <https://www.theaustralian.com.au/weekend-australian-magazine/true-disbeliever-senator-malcolm-roberts-crusade/news-story/0fd975d7b278dcb6a7d627ca2e62ee3e>.

Hodder, R 2011, 'Climate Conflict: Players and tactics in the greenhouse game', (Doctoral Thesis), University of Wollongong, Wollongong, <https://ro.uow.edu.au/cgi/viewcontent.cgi?article=4618&context=theses>.

Hudson, M 2018, 'Enacted Inertia: incumbent resistance to carbon pricing in Australia, 1989–2011', (Doctoral Thesis), University of Manchester, Manchester, <https://research.manchester.ac.uk/en/studentTheses/enacted-inertia-incumbent-resistance-to-carbon-pricing-in-austral>.

Jackson, J, 'A Response to Energy 2000 – A National Energy Policy Review', (Policy Submission), Bond Corporation.

'John Bonython Lecture Series 1984–2023, The Centre for Independent Studies, <https://www.cis.org.au/events/past-events/john-bonython-lectures-jbl/>.

Kelly, D 2019, *Political Troglodytes and Economic Lunatics: The hard right in Australia* (Black Inc., Melbourne.

Kelly, P 1992, *The End of Certainty: The story of the 1980s,* Allen & Unwin, Sydney.

Kemp, D & Van Doren, P 2022, 'Nuclear Power in the Context of Climate Change' Working Paper no. 68, Cato Institute, 26 July, <https://www.cato.org/working-paper/nuclear-power-context-climate-change>.

Lavoipierre, A 2016, 'Electricity boss mistakenly sent climate skeptic handbook to Australia's elite', *Triple J Hack*, ABC, 16 May, <https://www.abc.net.au/triplej/programs/hack/electricity-boss-mistakenly-sent-climate-skeptic-handbook/7418740>.

Lipsky, D 2023, *The Parrot and the Igloo*, WW Norton & Company, New York, extracted in 'How Sun Myung Moon "Digested the Scientists" and Fueled Climate-Change Denial' *Rolling Stone*, 20 August, <https://www.rollingstone.com/politics/politics-features/sun-myung-moon-unification-church-climate-change-denial-1234808552/>.

Morgan, P 2014, 'The Life and Times of Ray Evans', *Quadrant Online*, 19 June, <https://quadrant.org.au/opinion/qed/2014/06/life-times-ray-evans/>.

Oreskes, N & Conway, E 2011, *Merchants of Doubt,* Bloomsbury, London.

Pearse, G 2005, 'The Business Response to Climate Change: Case studies of Australian interest groups', (Doctoral Thesis), Australian National University, Canberra, <https://openresearch-repository.anu.edu.au/bitstream/1885/109792/4/b22560233-pearse_g.pdf>.

Ritchie, H & Roser, M 2020, 'CO_2 Emissions', Our World In Data, UK, June <https://ourworldindata.org/co2-emissions>.

Singer, SF 1989, *Global Climate Change: Human and natural influences,* Paragon House, Minnesota.

Shell 1986, 'Comments on discussion papers prepared for the Energy 2000 policy review' (Policy Submission), July.

Stafford, J, Bond, L & Daube, M 2009, *We are still not yet out of the Woods in WA: Western Australia and the international tobacco industry,* WA Tobacco Doucment Searching Program, Curtin Univeristy of Technology, Canberra, <chrome-extension://efaidnbmnnnibpcajpcglclefindmkaj/https://espace.curtin.edu.au/bitstream/handle/20.500.11937/37414/134730_17860_WATDSP%20monograph%20web%20version%20final.pdf?sequence=2&isAllowed=y >.

Victorian Government, 'Response to Energy 2000: A National Energy Policy Review', (Policy Submission).

'WA Businessman Helped Bankroll Abbott Fund' *ABC News,* 28 August 2003, <https://www.abc.net.au/news/2003-08-28/wa-businessman-helped-bankroll-abbott-fund/1471048>.

Walker, J, 'Freedom to Burn: Mining propaganda, fossil capital and the Australian neoliberals', in Slobodian, Q & Plehwe, D 2002, *Market Civilisations: Neoliberals East and South,* Zone Books, Princeton University Press, New Jersey. <https://www.academia.edu/81932468/Freedom_to_Burn_Mining_Propaganda_Fossil_Capital_and_the_Australian_Neoliberals_DRAFT_?f_ri=93298>.

Weidenbach, K 2014, *Blue Flames, Black Gold: The story of Santos,* Santos Pty Ltd, Adelaide, pp. 29–34, 35–36.

Wilkinson, M 2020, *The Carbon Club,* Allen & Unwin, Sydeny, pp. 36–43.

Wilkinson, R 2010, *Knights, Knaves And Dragons: 50 Years Inside APPEA and Australia's Oil and Gas Politics,* Media Dynamics, Winton, Queensland, pp. 230–231.

Chapter 11

Augustine, G & Soppe, B 2023, 'Exodus from Fossil Fuels: Constructing dystopias that prompt prefigurative work', (Working Paper), 17th Organization Studies Workshop, Athens.

Dunlop, I & Evans, R 2009, 'The Theology of Climate Change' (Debate), The Sydney Institute, 30 June.

Halttunen, K, Slade, R & Staffel, I 2022, '"We Don't Want To Be the Bad Guys": Oil industry's sensemaking of the sustainability transition paradox', *Energy Research & Social Science,* vol. 92, October, <https://www.sciencedirect.com/science/article/pii/S2214629622003036>.

Heede, R 2014, 'Carbon Majors: Accounting for carbon and methane emissions from 1854–2010 (Methods and Results Report), Climate Justice Programme, Sydney & Greenpeace International, Amsterdam, <https://climateaccountability.org/pdf/MRR%209.1%20Apr14R.pdf>.

International Energy Agency 2019, 'Energy Section Employment by Region', *World Energy Employment: Overview*, (Report), August, <https://www.iea.org/reports/world-energy-employment/overview>.

McKie, R 2018, 'Climate Change Counter Movement Neutralization Techniques: A typology to examine the climate change counter movement', *Sociological Inquiry*, vol. xx, no. x, Alpha Kappa Delta: The International Sociology Honor Society, New York, <https://cssn.org/wp-content/uploads/2020/12/soin_12246_Rev_EV.pdf>.

Morton, A 2009, 'The Sceptic's Shadow of Doubt', *The Sydney Morning Herald*, 2 May, <https://www.smh.com.au/national/the-sceptics-shadow-of-doubt-20090501-aqa1.html>.

Ritchie, H & Roser, M 2020, 'CO_2 Emissions', Our World In Data, UK, June <https://ourworldindata.org/co2-emissions>.

Robert, J 2021, 'IEEFA Update: Santos won't solve the problem of Barossa LNG with carbon capture and storage', (Report), Institute for Energy Economics and Financial Analysis, 20 October, <https://ieefa.org/articles/ieefa-update-santos-wont-solve-problem-barossa-lng-carbon-capture-and-storage>.

Santos, 'Acquisition of ConocoPhillips' Northern Australia Interests', (Media Release), 14 October 2019, <https://www.santos.com/news/acquisition-of-conocophillips-northern-australia-interests/>.

Taylor, M & Watts, J 2019, 'Revealed: the 20 firms behind a third of all carbon emissions' *The Guardian*, 9 October, <https://www.theguardian.com/environment/2019/oct/09/revealed-20-firms-third-carbon-emissions>.

Chapter 12

'Albert Namatjira & WG Walkley looking at cigar', 1947–1950, (Photograph), PIC/7569-Albert Namitjira photograph collection, <https://nla.gov.au/nla.obj-146941064/view>.

'Albert Namatjira standing next to his loaded new truck along with Keith Namatjira, Frank Clune and WG Walkley', 1947–1950, (Photograph), PIC/7569-Albert Namitjira photograph collection, < https://trove.nla.gov.au/work/16694841?keyword=albert%20namatjira%20william%20walkley>.

'Albert Namatjira wishing goodbye to WG Walkley outside Romano's Restaurant Sydney', 1947–1950, (Photograph), PIC/7569-Albert Namatjira photograph collection, <https://trove.nla.gov.au/work/16694934?keyword=albert%20namatjira%20william%20walkley>.

Bridges, A & Trilling, J 2022, 'Tim Winton uses Perth Festival address to denounce fossil fuel reliance, arts sponsorship', *ABC News*, 28 February, <https://www.abc.net.au/news/2022-02-28/tim-winton-perth-festival-fossil-fuels-sponsorship-woodside/100868632>.

Burke, K 2022, '"It makes us chumps": Tim Winton speaks out against fossil fuel sponsorship of Perth festival', *The Guardian*, 28 February, <https://www.theguardian.com/culture/2022/feb/28/it-makes-us-chumps-tim-winton-speaks-out-against-fossil-fuel-sponsorship-of-perth-festival>.

Centre for International Law, 'Smoke', *Smoke & Fumes*, <https://www.smokeandfumes.org/smoke>.

Clements, A, Maddocks, F, Lewis, J, Molleson, K, Service, T, Jeal, E & Ashley T 2009, 'The Best Classical Music Works of the 21st Century', *The Guardian*, 13 September, <https://www.

theguardian.com/music/2019/sep/12/best-classical-music-works-of-the-21st-century>.

de Kruijff, P 2022, '"Smouldering dumpster fire": Literary giant Tim Winton pokes the oil and gas bear', *The Sydney Morning Herald*, 28 February, <https://www.abc.net.au/news/2022-02-28/tim-winton-perth-festival-fossil-fuels-sponsorship-woodside/100868632>.

Disrupt Burrup Hub 2023, 'Artists deface iconic Aussie painting to protest Woodside's Burrup Hub', (Facebook Video 3:04), 19 January, <https://www.facebook.com/disruptburruphub/videos/1325072355003547>.

Faruqi, O 2023, 'White supremacy and the name synonymous with the best journalism in this country', *The Sydney Morning Herald*, 2 September, <https://www.smh.com.au/national/the-racist-past-of-the-oil-baron-who-set-up-australia-s-top-journalism-award-20230901-p5e18i.html>.

Amussen, CH 1975, Interview with Hill, JW: 'Original Copy of Golden Interview', (Transcript), p. 53, Univeristy of San Francisco, California, <https://www.industrydocuments.ucsf.edu/docs/#id=xyln0042>.

Joshi, K 2023, 'The Walkley's Walkout' (Blog), 30 August, <https://ketanjoshi.co/2023/08/30/the-walkleys-walkout/>.

Kurmelovs, R 2021, 'Fossil fuel advertising in sport "the new cigarette sponsorship", ex-Wallabies captain David Pocock says' *The Guardian*, 10 November, <https://www.theguardian.com/business/2021/nov/10/fossil-fuel-advertising-in-sport-the-new-cigarette-sponsorship-says-ex-wallabies-captain-david-pocock>.

Kurmelovs, R 2021, 'WA Nippers parents speak out against Woodside Energy sponsorship deal' *The Guardian*, 22 November, <https://www.theguardian.com/australia-news/2021/nov/22/wa-nippers-parents-speak-out-against-woodside-energy-sponsorship-deal>.

Kurmelovs, R 2022, 'Perth festival production climate change work criticised for "farcical" fossil fuel sponsorship' *The Guardian*, 11 February, <https://www.theguardian.com/culture/2022/feb/11/perth-festival-production-of-climate-change-work-criticised-for-farcical-fossil-fuel-sponsorship>.

Kurmelovs, R 2022, 'Fossil fuel interests revealed to have sponsored more than 500 Australian community organisations' *The Guardian*, 27 December, <https://www.theguardian.com/environment/2022/dec/27/fossil-fuel-interests-revealed-to-have-signed-more-than-500-sponsorship-deals-with-australian-bodies>.

Noble, B 2023, 'The awkward snub of climate change by Ampol-sponsored Walkley Awards', *Mumbrella*, 22 May, <https://mumbrella.com.au/the-awkward-snub-of-climate-change-by-ampol-sponsored-walkley-awards-788011>.

O'Connell, J, '1927 Romano's restaurant opens in Sydney', (Blog), <https://australianfoodtimeline.com.au/romanos-restaurant-history/>.

Oreskes, N & Conway, E 2011, *Merchants of Doubt*, Bloomsbury, New York.

'Richard W Darrow Dies at 60; Top Officer of Hil & Knowlton', *The New York Times* (Obituary) 21 March 1976, <https://www.nytimes.com/1976/03/21/archives/richard-w-darrow-dies-at-60-top-officer-of-hill-knowlton.html>.

Stafford, J, Bond, L & Daube, M 2009, *We are still not yet out of the Woods in WA: Western Australia and the international tobacco industry*, WA Tobacco Doucment Searching Program, Curtin Univeristy of Technology, Canberra, <chrome-extension://efaidnbmnnnibpcajpcglclefindmkaj/https://espace.curtin.edu.au/bitstream/handle/2

0.500.11937/37414/134730_17860_WATDSP%20monograph%20web%20version%20
final.pdf?sequence=2&isAllowed=y >

'The origin of Australia's most prestigious journalism awards', (Statement), *Walkley Foundation*, 16 November 2022, <https://www.walkleys.com/the-origin-of-australias-most-prestigious-journalism-awards/>.

'The Western Australian Government's Attempts to Pass Legislation to Ban Tobacco Promotion' (Watching Brief,) Tobacco Industry Archives, Tobacco Institute of Australia, <https://www.industrydocuments.ucsf.edu/tobacco/docs/#id=zrxj0136>.

Truth Tobacco Industry Documents, (Letter thought to be from former Medical Director of Standard Oil, JP Holt, recommending scientists from Standard Oil for the Tobacco Industry Scientific Advisory Board), <https://www.industrydocuments.ucsf.edu/tobacco/docs/#id=xywk0012>.

Truth Tobacco Industry Documents, (Memo regarding 'Tobacco Industry Meeting, New York' and hiring of Hill&Knowlton), 14 December 1953, < https://www.industrydocuments.ucsf.edu/tobacco/docs/#id=ghxd0040>.

Truth Tobacco Industry Documents, (Memo to Hill&Knowlton staff regarding latest news and information, including activities of Darrow, R, in Western Australia), 23 April 1954, pp. 3–4, <https://www.industrydocuments.ucsf.edu/docs/#id=lznl0042>.

Truth Tobacco Industry Documents, (Memo from Darrow, R to Hartnett, T, regarding excerpts from constitutions), 13 December 1954, <https://www.industrydocuments.ucsf.edu/tobacco/docs/#id=hnln0042>.

Truth Tobacco Industry Documents, (Speaking notes of Reynolds, RJ, address regarding importance of special events to tobacco), p. 11, <https://www.industrydocuments.ucsf.edu/tobacco/docs/#id=jrkm0084>.

Winton, T 2022. 'Closing Address' Perth Writers Festival, (Speech), 27 February, <https://www.marineconservation.org.au/gaslit-edited-excerpt-from-our-patron-tim-wintons-closing-address-at-the-perth-writers-festival-27-2-22/>.

''53 Experiment Found Cancer Link', *The New York Times*, 16 July 1994, <https://www.nytimes.com/1994/06/16/us/53-experiment-found-cancer-link.html#:~:text=Ernst%20Wynder%20and%20his%20colleagues,they%20had%20been%20too%20brief.>.

Chapter 13

ACCR 2023, 'Santos progresses marginal fossil fuel project under guise of energy security, CEO's pay packet the only clear winner', (Media Release), Australian Centre for Corporate Responsibility, 14 February, <https://www.accr.org.au/news/santos-progresses-marginal-fossil-fuel-project-under-guise-of-energy-security-ceo%E2%80%99s-pay-packet-the-only-clear-winner/#:~:text=Santos'%20CEO%20Growth%20Projects%20Incentive,Santos'%20major%20growth%20projects%E2%80%9D>.

Barlass, T 2022, 'Santos salt of the earth could be revenue stream for Narrabri mayor', *The Sydney Morning Herald*, 20 August, <https://www.smh.com.au/national/santos-salt-of-the-earth-could-be-revenue-stream-for-narrabri-mayor-20220817-p5bak6.html>.

Clark, H 2023, 'NOPSEMA raid lands Santos with Barossa directions post court case loss', *Energy News Bulletin*, 24 January, <https://www. energynewsbulletin.net/finance-legal/news/1446970/nopsema-raid-lands-santos-with-barossa-directions-post-court-case-loss>.

Cox, L 2023, '"We Will Win": Tiwi Islanders draw a line in the sand against Santos gas project and "white fella rule"', *The Guardian*, 22 June, <https://www.theguardian.com/australia-news/2023/jun/21/we-will-win-tiwi-islanders-draw-a-line-in-the-sand-against-santos-gas-project-and-white-fella-rule>.

'"Enough is Enough": Tiwi Islander takes Australian government to court over the Barossa Gas Project as drilling about to begin', (Media Release), Climate Media Centre, 7 June 2022.

Grounds, E 2022, 'The Gomeroi people have fought Santos' Narrabri Gas Project for a decade. They hope a novel climate change argument could help them win', *Triple J Hack*, ABC, 13 September, <https://www.abc.net.au/news/2022-09-13/gomeroi-fight-santos-narrabri-gas-project-climate-change/101428862>.

NOPSEMA, 'General Direction – s 574', (Direction no. 1898), 13 January 2023, <https://www.nopsema.gov.au/sites/default/files/documents/General%20Direction%201898.pdf>.

'Remuneration Report', Santos 2022 Annual Report, Santos PTY Ltd, Adelaide, <https://www.santos.com/wp-content/uploads/2023/02/2022-Annual-Report.pdf>.

Santos 2022, 'Federal Court Decision' (Media Release), Santos Pty Ltd, Adelaide, 21 September, <https://www.santos.com/news/federal-court-decision/>.

Santos NSW Pty Ltd and Another v. Gomeroi People and Another (2022), National Native Title Tribunal of Australia, 74, <http://www.nntt.gov.au/searchRegApps/FutureActs/FA%20Determination%20Documents/NSW/FutureActsDeterminations/2022/December/NF2021_0003-0006%2019122022.pdf>.

Chapter 14

APPEA 2022, 'APPEA requests urgent meeting with PM as more damaging intervention emerges', (Media Release), APPEA, 11 December.

'APPEA appoints new chief executive', *Energy Source & Distribution*, 5 February 2019, <https://esdnews.com.au/appea-appoints-new-chief-executive/>.

Aston, H 2015, 'Union official attempting to unseat Labor MP criticised over satirical female advice book', *The Sydney Morning Herald*, 20 November, <https://www.smh.com.au/politics/federal/union-official-attempting-to-unseat-labor-mp-criticised-over-satirical-female-advice-book-20151119-gl38yh.html>.

Aston, H 2015, 'Labor hopeful Misha Zelinsky apologise over "joke" advice book for women', *The Sydney Morning Herald*, 20 November, <https://www.smh.com.au/politics/federal/labor-hopeful-misha-zelinsky-apologises-over-joke-advice-book-for-women-20151120-gl3zu7.html>.

Australian Energy Producers 2022, 'Gas market intervention a warning for Australian industry', (Media Release), 15 December, <https://energyproducers.au/all_news/media-release-gas-market-intervention-a-warning-for-australian-industry/>.

Australian Energy Producers 2022, 'Investment risk of intervention shown in Senex

Energy pause on $1 billion expansion', (Media Release), 22 December, <https://energyproducers.au/all_news/media-release-investment-risk-of-intervention-shown-in-senex-energy-pause-on-1-billion-expansion/>.

Australian Energy Producers 2023, 'Gas sector to deliver $16 billion to government as contribution almost triples', (Media Release), 21 April, <https://energyproducers.au/all_news/media-release-gas-sector-to-deliver-16-billion-to-governments-as-contribution-almost-triples/>.

Australian Energy Producers, 2023 'Beetaloo green light ushers in economic prosperity and east coast energy security' (Media Release), 3 May, < https://energyproducers.au/all_news/media-release-beetaloo-green-light-ushers-in-economic-prosperity-and-east-coast-energy-security/>.

Australian Energy Producers 2023, 'Future role of gas and net zero tech headline federal budget underpinned by oil and gas contribution', (Media Release), 9 May, <https://energyproducers.au/all_news/media-release-future-role-of-gas-and-net-zero-tech-headline-federal-budget-underpinned-by-oil-and-gas-contribution/>.

Barnett, A & Bright, S 2023, 'Rishi Sunak boasts that oil funded think tank "helped us draft" crackdown on climate protests', *DeSmog*, 29 June, <https://www.desmog.com/2023/06/29/rishi-sunak-boasts-that-oil-funded-think-tank-helped-us-draft-crackdown-on-climate-protests/>.

Bashan, Y 2022, 'Gas lobby group APPEA a little late to the party,' *The Australian*, 24 April, <https://www.theaustralian.com.au/business/margin-call/gas-lobby-group-appea-a-little-late-to-the-party/news-story/da50b595d6d99501537e8340b9382c3f>.

Bashan, Y 2023, 'Alex Turnbull hires Bret Walker SC for appeal against judgment in Russell Pillemer case' *The Australian*, 1 February 2023, <https://www.theaustralian.com.au/business/margin-call/loss-staff-and-corporate-knowledge-from-the-upper-levels-of-appeas-management/news-story/a1adef3d25aeea7b816927c0f8c7f842>.

Bucci, N 2022, 'Blockade Australia activists say police carrying out surveillance refused to identify themselves,' *The Guardian*, 21 June, <https://www.theguardian.com/australia-news/2022/jun/21/blockade-australia-activists-say-police-carrying-out-surveillance-refused-to-identify-themselves>.

Horton, H 2022, 'Thinktank that brief against XR given $30k by ExxonMobil in 2017', *The Guardian*, 15 June, <https://www.theguardian.com/environment/2022/jun/15/thinktank-that-briefed-against-xr-given-30k-by-exxon-mobil-in-2017>.

'IGU Position on Climate Change and the Future Role of Gas', (Policy Statement), International Gas Union, 13 May 2021.

International Gas Union 2021, Executive Committee Meeting 'COP26 Advocacy', (Video Conference).

Kurmelovs, R 2021, 'Chalk paint and police raids: why climate activists are under fire', *The Guardian*, 4 December, <https://www.theguardian.com/environment/2021/dec/04/chalk-paint-and-police-raids-why-climate-activists-are-under-fire>.

Kurmelovs, R 2022, 'Fireproof Australia: who are the radical Extinction Rebellion splinter Group?', *The Guardian*, 10 April, <https://www.theguardian.com/environment/2022/apr/10/fireproof-australia-who-are-the-radical-extinction-rebellion-splinter-group>.

Kurmelovs, R 2022, 'Strict anti-protest laws may have encouraged mining conference to move from Melbourne to Sydney' *The Guardian*, 29 October, <https://www.theguardian.com/business/2022/oct/29/strict-anti-protest-laws-may-have-encouraged-mining-conference-to-move-from-melbourne-to-sydney>.

Kurmelovs, R 2023, 'Lismore one year after the floods,' *The Saturday Paper*, 4 March, <https://www.thesaturdaypaper.com.au/news/economy/2023/03/04/lismore-one-year-after-the-floods>.

Kurmelovs, R 2023, 'Climate activist Violet CoCo and protest laws,' *The Saturday Paper*, 18 March, <https://www.thesaturdaypaper.com.au/news/environment/2023/03/18/climate-activist-violet-coco-and-protest-laws>.

Lakhani, N 2022, 'Revealed: rightwing US lobbyists help craft slew of anti-protest fossil fuel bills' *The Guardian*, 15 September, <https://www.theguardian.com/us-news/2022/sep/14/rightwing-lobbyists-at-heart-of-anti-protest-bills-in-republican-states>.

Norington, B 2022, 'Labor preselection hopeful Misha Zelinsky forced to quit', *The Australian*, 10 February, <https://www.theaustralian.com.au/nation/politics/labor-preselection-hopeful-misha-zelinsky-forced-to-quit/news-story/d927742223fc03be7329e3c522adf16f>.

Parkes-Hupton, H 2022, 'NSW parliament passes new laws bringing harsher penalties on protesters' *ABC* News, 1 April, <https://www.abc.net.au/news/2022-04-01/nsw-new-protest-laws-target-major-economic-disruption/100960746>.

Police, Crime, Sentencing and Courts Act 2022 (UK).

Rose, T & McGowan, M 2023, 'NSW premier describes jailing of climate activist Deanna "Violet" Coco as "pleasing to see"', *The Guardian*, 5 December, <https://www.theguardian.com/world/2022/dec/05/deanna-violet-coco-jailed-climate-activist-protester-sydney-harbour-bridge-nsw-premier>.

Sutton, C 2020, 'Roxy Jacenko flees court in tears after magistrate's "damning" judgment', *News.com.au*, 28 February, <https://www.news.com.au/national/nsw-act/courts-law/shock-claim-from-man-accused-of-defacing-roxy-jacenkos-sydney-office/news-story/5f83fbaa06e15ca75b26d07e8adaa71c>.

'The Policing Bill – What happened, and what now?', *Liberty*, 29 April, <https://www.libertyhumanrights.org.uk/issue/the-policing-bill-what-happened-and-what-now/>.

Thomas, K 2021, 'NRL star Josh Addo-Carr avoids conviction after unauthorised firearm use during NSW camping trip' *ABC News*, 5 March, <https://www.abc.net.au/news/2021-03-05/nrl-star-josh-addo-carr-avoids-conviction-over-firearm-charge/13215796>.

Turnbull, T 2022, 'Violet Coco: Climate activist's jailing ignites row in Australia' *BBC*, 8 December, <https://www.bbc.com/news/world-australia-63883430>.

Weidenbach, K 2014, *Blue Flames, Black Gold: the story of Santos*, Santos Pty Ltd, Adelaide.

Western Australia Parliament, Transcript of Evidence – Session One; Western Australia Police Force, Standing Committee on Estimates and Financial Operations, 29 June 2023, p. 28, <https://www.parliament.wa.gov.au/Parliament/commit.nsf/(Evidence+Lookup+by+Com+ID)/02470411170C62AC482589CE0029E6A3/$file/ef.be24.230629.tro.001.cb.pdf>.

Wilson, T & Walton, R 2019, 'Extremism Rebellion: a review of ideaology and tactics', (Report), *Policy Exchange*, 16 July 2019, <https://policyexchange.org.uk/publication/extremism-rebellion/>.

Zelinsky, M 2023, 'Responding to the European gas crisis – getting the politics, policy and pressures right', (Speech), Australian Domestic Gas Outlook, 22 March.

Chapter 15

Bermingham, K 2022, 'Premier Peter Malinauskas vows no conflicts over brother Rob's Santos job', *The Advertiser*, 24 May, <https://www.adelaidenow.com.au/news/south-australia/premier-peter-malinauskas-vows-no-conflicts-over-brother-robs-santos-job/news-story/a58f8d98cc44b90bc43bf98899a3e390>.

Harvey, F 2023, 'World likely to breach 1.5C climate threshold by 2027, scientists warn', *The Guardian*, 17 May, <https://www.theguardian.com/environment/2023/may/17/global-heating-climate-crisis-record-temperatures-wmo-research>.

Hoffman, AJ & Ely, DM 2022, 'Time to put the fossil-fuel industry into hospice', *Stanford Social Innovation Review*, <https://ssir.org/articles/entry/time_to_put_the_fossil_fuel_industry_into_hospice>.

Kelsall, T 2023, '"Iron Fist": Govt blasted as protest law passes', *InDaily*, 31 May, <https://indaily.com.au/news/2023/05/31/iron-fist-govt-blasted-as-protest-law-passes/>.

Kurmelovs, R 2022, 'South Australia won't supply gas to NSW during shortages, minister warns', *The Guardian*, 21 October, <https://www.theguardian.com/australia-news/2022/oct/21/south-australia-wont-supply-gas-to-nsw-during-shortages-minister-warns>.

Kurmelovs, R 2023, 'South Australia tells gas industry the state is "at your disposal"', *The Guardian*, 15 May, <https://www.theguardian.com/environment/2023/may/16/south-australia-gas-industry-appea-national-conference-2023>.

Kurmelovs, R 2023, 'South Australia rushes through anti-protest laws as activists rally outside oil and gas conference' *The Guardian*, 18 May, <https://www.theguardian.com/australia-news/2023/may/18/south-australia-anti-protest-laws-activists-rally-oil-gas-appea-conference-adelaide>.

Penberthy, D 2023, 'Left dangling in a climate protest crackdown' *The Australian*, 27 March, <https://www.theaustralian.com.au/inquirer/left-dangling-in-a-climate-protest-crackdown/news-story/821f3579c6a41a497b2f6c87e290380a>.

Chapter 16

Energy Industries Council 2019, 'Q&A: View from the top – HE Dr Sultan Ahmed Al Jaber', *EnergyFocus*, 31 October, <https://energyfocus.the-eic.com/eic/qa-view-top-he-dr-sultan-ahmed-al-jaber>.

Kurmelovs, R 2023, 'Inside the COP28 talks', *The Saturday Paper*, 2 December, <https://www.thesaturdaypaper.com.au/news/environment/2023/12/02/inside-the-cop28-talks>.

Kurmelovs, R 2023, 'How COP28 fell short', *The Saturday Paper*, 16 December, <https://www.thesaturdaypaper.com.au/news/environment/2023/12/16/how-cop28-fell-short>.

Raval, A & Kerr, S 2021, 'Adnoc defies retreat from oil with push to pump up output', *Financial Times*, 18 January, <https://www.ft.com/content/b13efad3-f494-46a8-842c-ccf7124ad43d>.

Stockton, B & Westervelt, A 2023, 'Inside the campaign that put an oil boss in charge of a climate summit', *Centre for Climate Reporting*, 25 October, <https://climate-reporting.org/al-jaber-cop28-uae-oil/>.

Reuters 2023, 'Woodside and Santos in talks over merger to create A$80bn energy behemoth', *The Guardian*, 7 December, <https://www.theguardian.com/australia-news/2023/dec/07/woodside-and-santos-in-talks-over-merger-to-create-a80bn-energy-behemoth>.